Gamer's Gizmos and Gadgets

OTHER BOOKS BY PHILIP GARNER

Philip Garner's Better Living Catalog
Utopia . . . or Bust!

GARNER'S GIZMOS AND GADGETS

Philip Garner

A Perigee Book

Perigee Books
are published by
The Putnam Publishing Group
200 Madison Avenue
New York, NY 10016

Published simultaneously in Canada by
General Publishing Co. Limited, Toronto

Library of Congress Cataloging-in-Publication Data

Garner, Philip.
Garner's gizmos and gadgets.

1. Commercial products—United States—
Caricatures and cartoons. 2. American
wit and humor, Pictorial. I. Title.
II. Title: Gizmos and gadgets.
NC1429.G28A4 1986 741.5'973 86-30270
ISBN 0-399-51343-4

Printed in the United States of America
1 2 3 4 5 6 7 8 9 10

INTRODUCTION

Perhaps my reputation precedes me as the creator of useful items for areas of need overlooked by the main thrust of the industrial age.

Having fostered such indispensable contrivances as the fold-away Murphy Grand Piano, the Outboard Motor-Cycle (you picture it), and the Pillow Hat which enables the wearer to nap anywhere (leaning against a tree, for instance), I have raised a few eyebrows in my time. One question frequently asked is: "Where do you get your ideas from?" My usual answer: "I get them from Little Phil, my brain," is more often taken as a "smart remark" than a satisfactory reply. I have therefore decided to bare all, come clean, and, in a nutshell, reveal for the first time ever the most intimate and innermost perceptions lurking in the darkest recesses of my mind's eye.

The following pages contain highlights from my personal sketchidea file, a dusty manuscript upon which eyes other than my own (Little Phil's own) have never been laid.

As few of the items illustrated herein have even reached the pre-production-prototype stage, I ask you, the untold masses, as a personal favor to me, to keep what you see here under your hat.

Thank you,
Philip Garner

PERSONAL SATELLITE TV

ELECTRONIC AUDIO-ZIPPER
warns of incomplete closure

"LIBRARY" ADHESIVE-BACK PAPER

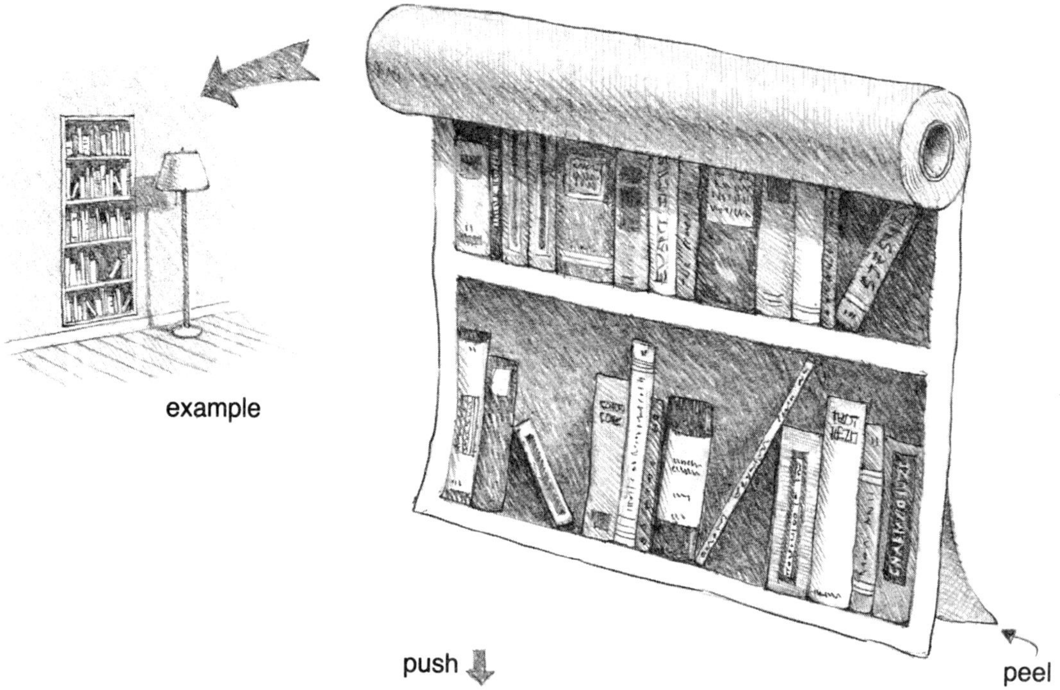
example
peel

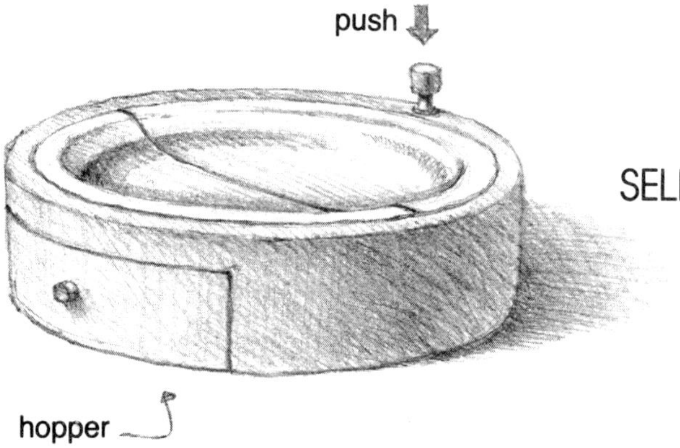
push
hopper

SELF-CLEANING PLATE

GAS-POWERED TOOTHBRUSH

(for electrical blackouts)

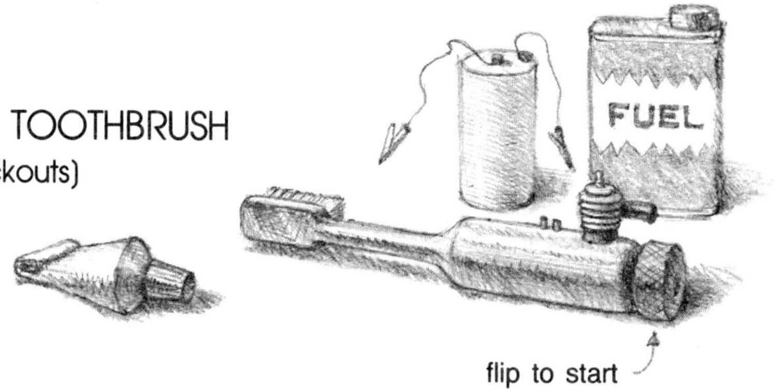
FUEL
flip to start

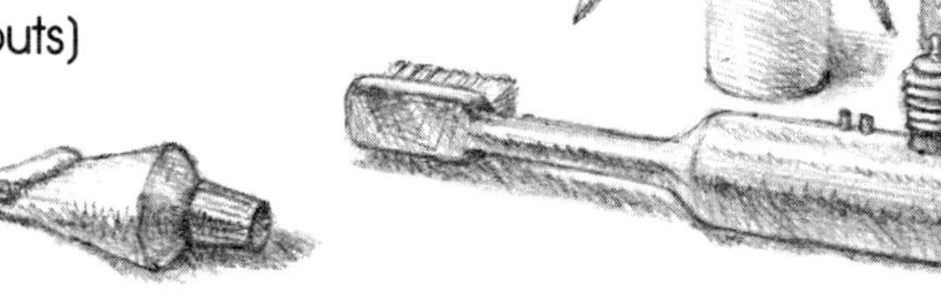

BACK-DRY CHAIR

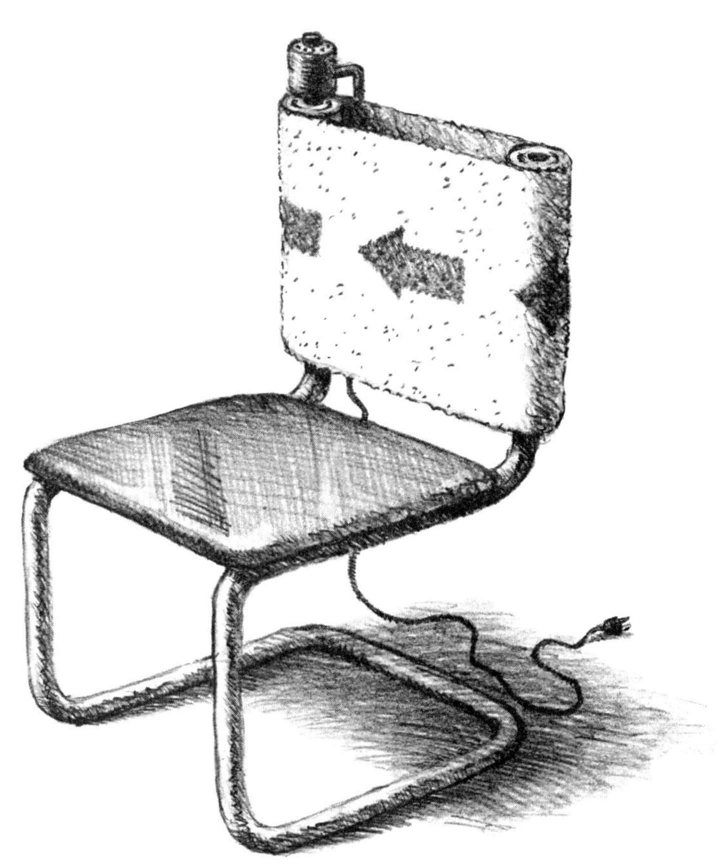

BACON WATCHBAND

ELECTRIC "BLANK-IT" THOUGHT REMOVER

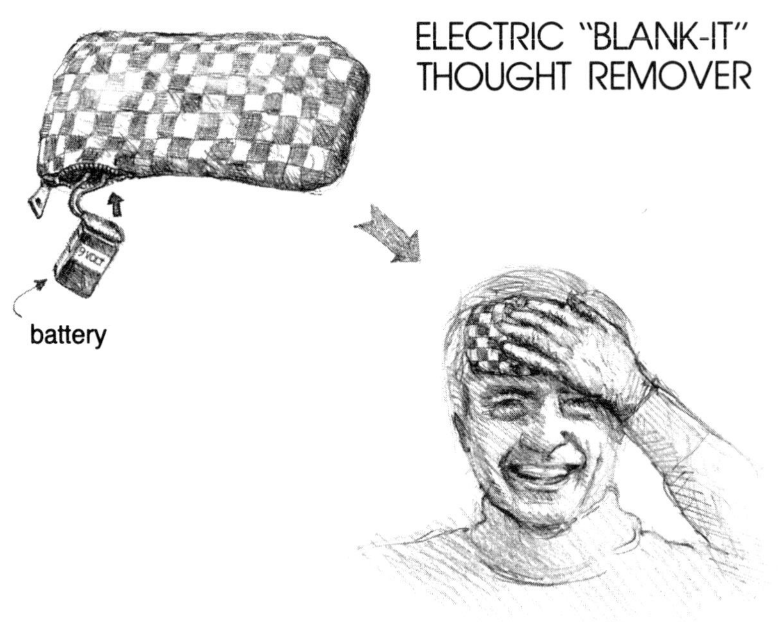

X-TRA BRAIN

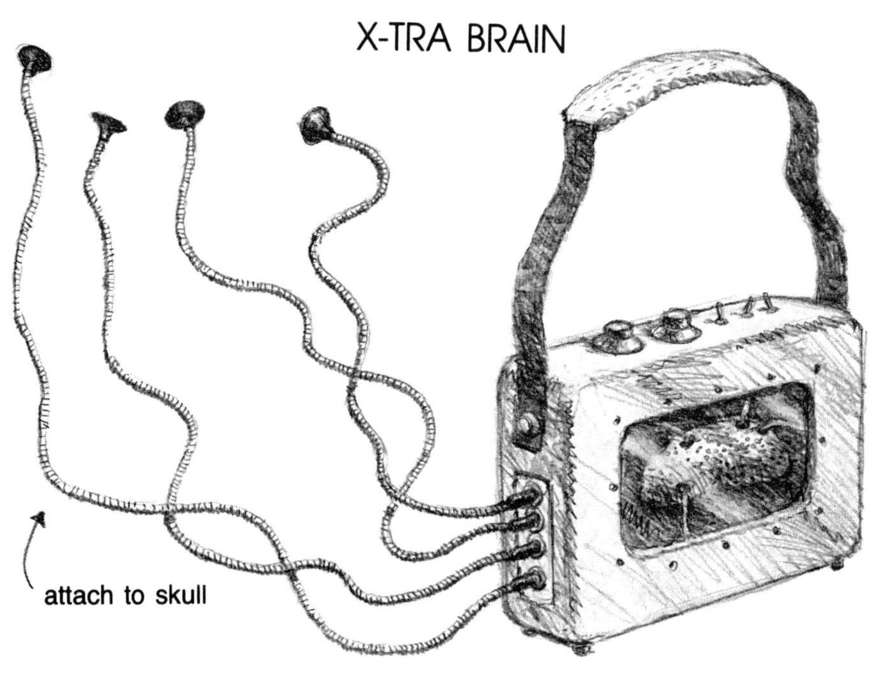

CYCLOPS GLASSES

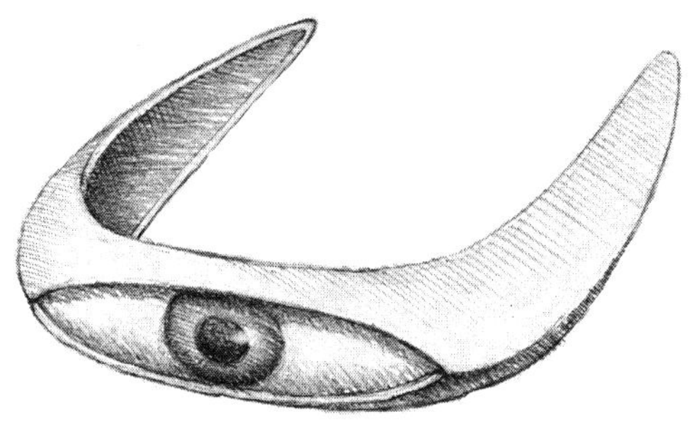

ESKIMO PIE COMPACT

SHOOT-A-CARD

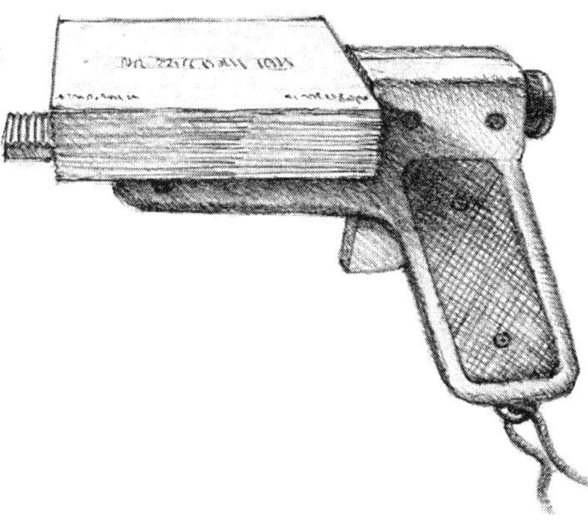

SUPER BROOM

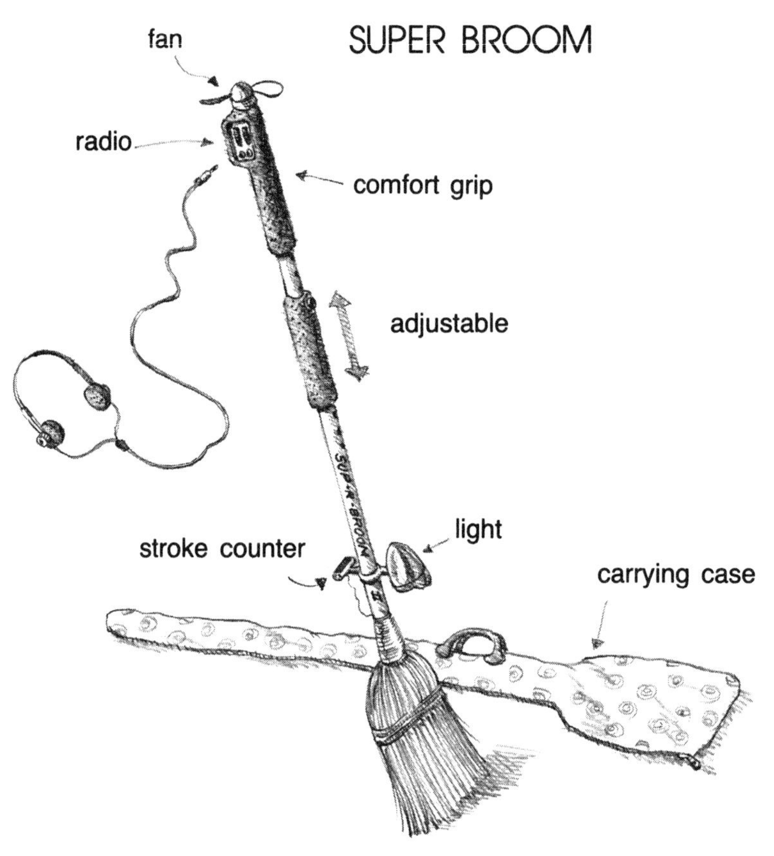

FOUNTAIN TIE

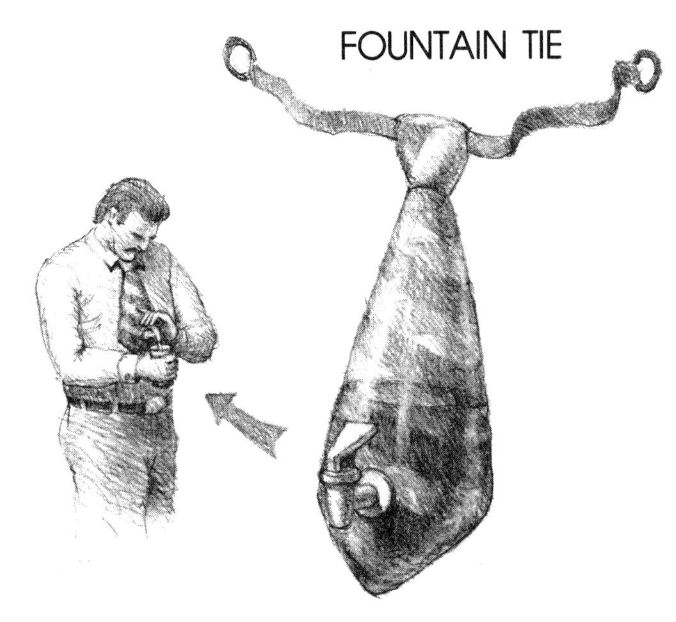

SIAMESE SCHOOL DESKS

POTTED TV

MOVING FRIEZE

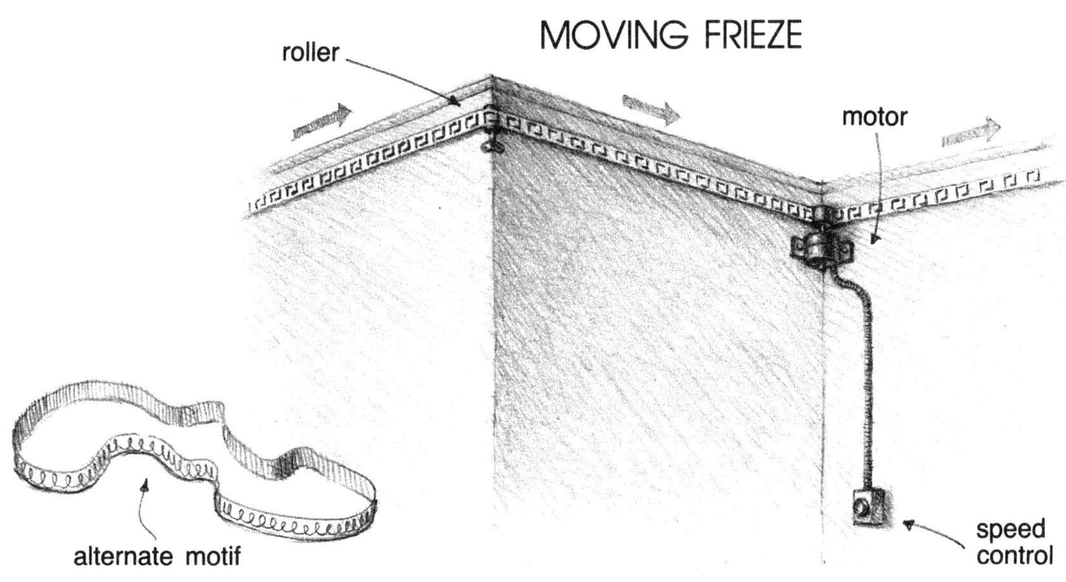

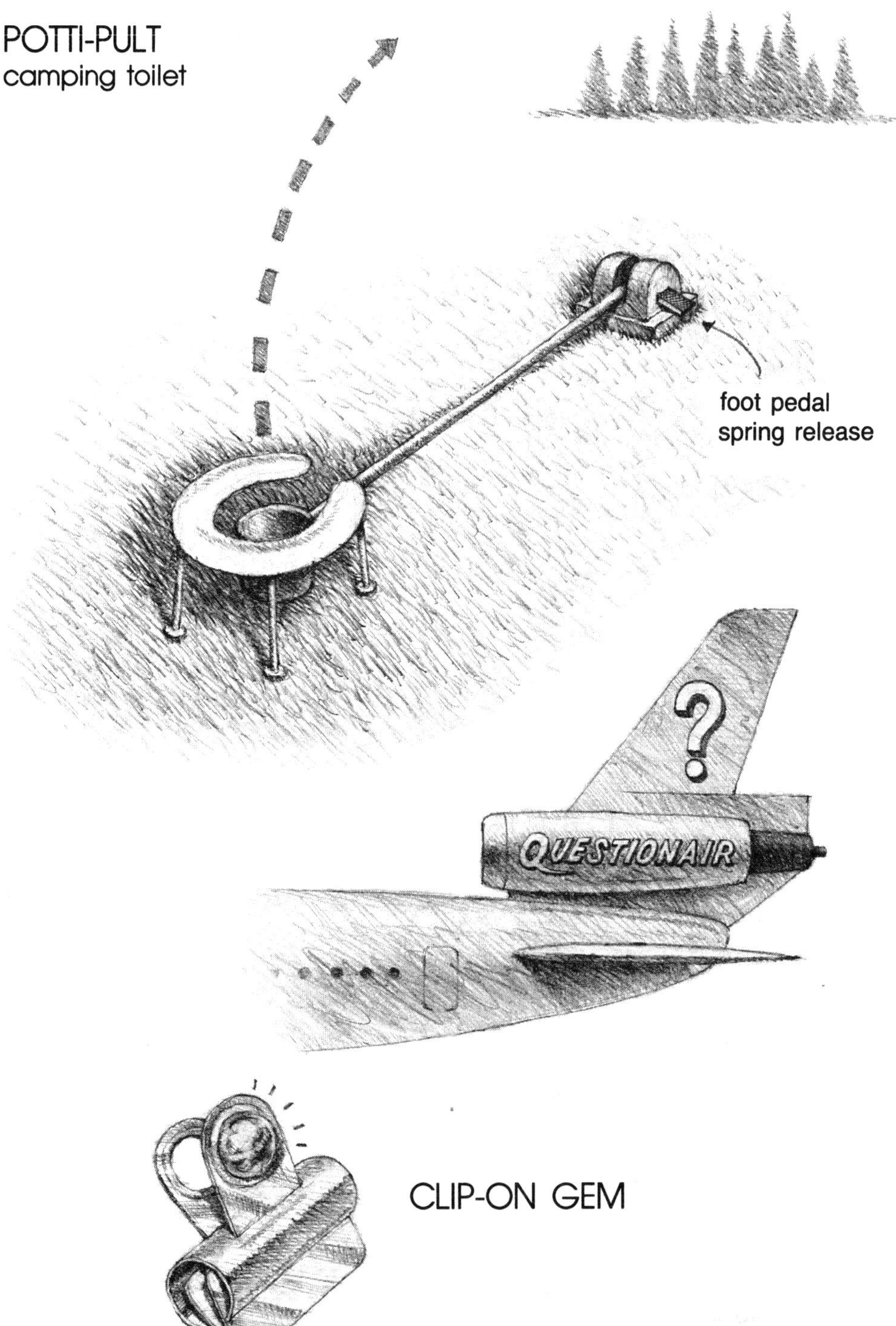
POTTI-PULT
camping toilet
foot pedal
spring release
?
QUESTIONAIR
CLIP-ON GEM

"SPRAVY"
AEROSOL GRAVY

STREET PERFORMER'S STARTER KIT

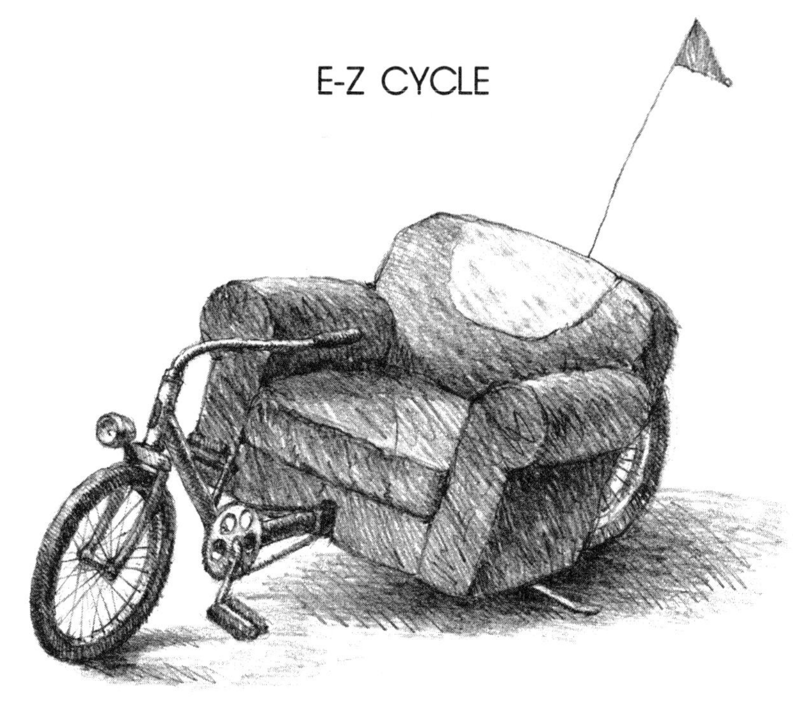

E-Z CYCLE

SOFA-SLIDE

BOARDED-UP TELEVISION
JET MELON
X-TEND-A-BILL CAP

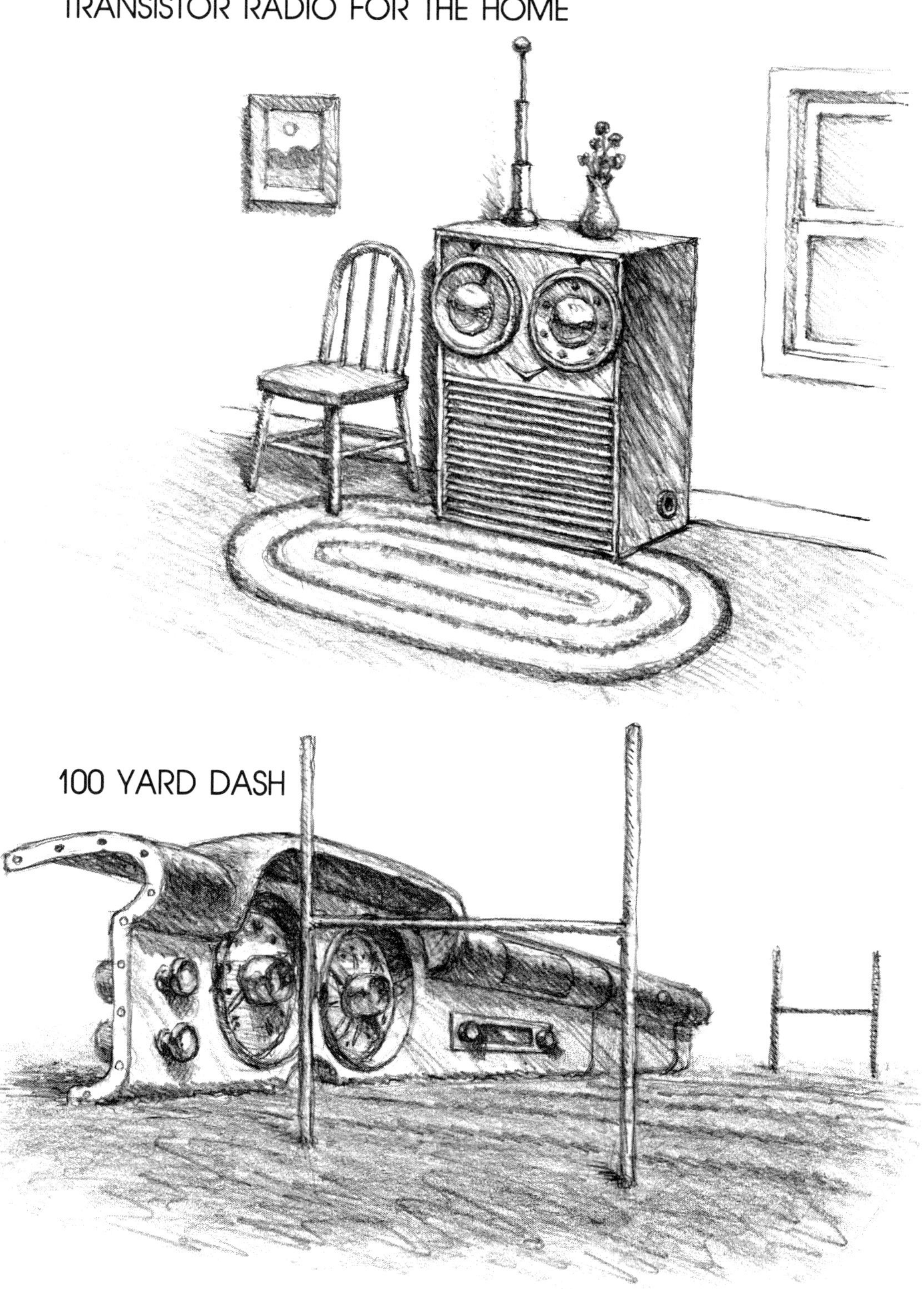
TRANSISTOR RADIO FOR THE HOME
100 YARD DASH

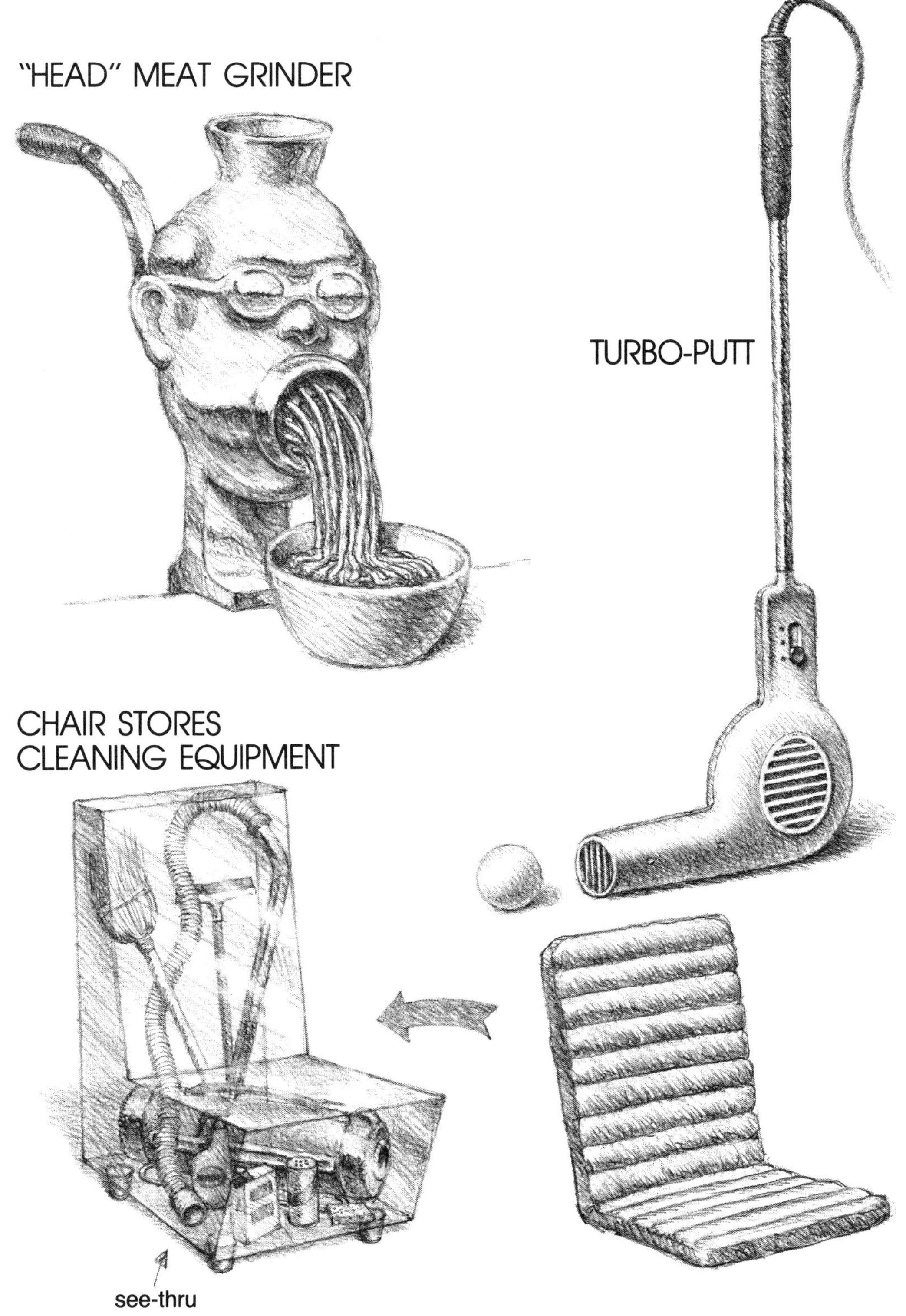
"HEAD" MEAT GRINDER
TURBO-PUTT
CHAIR STORES
CLEANING EQUIPMENT
see-thru

REAR-VIEW HAT

LUNCH-O-MATIC

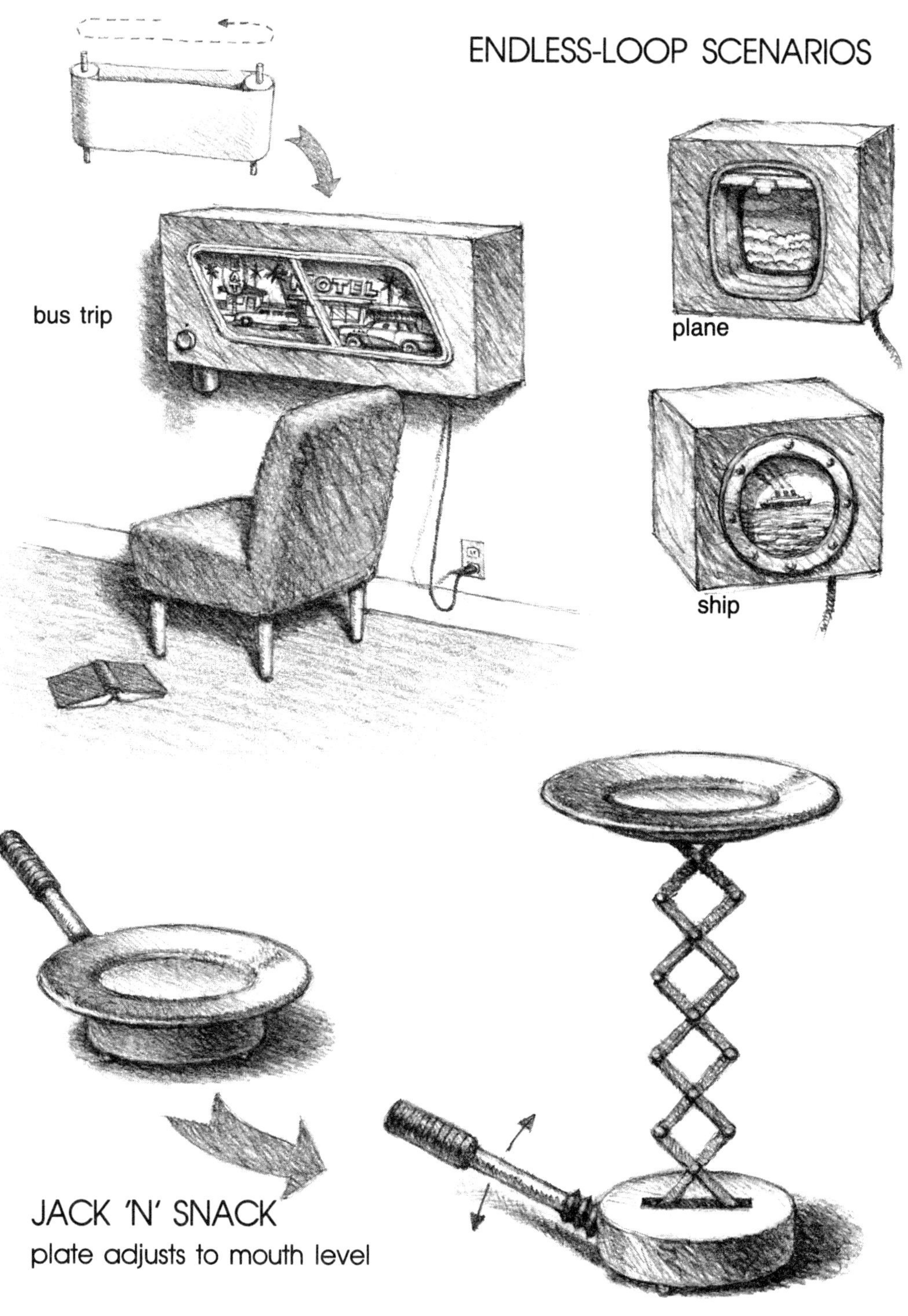
ENDLESS-LOOP SCENARIOS
bus trip
plane
ship
JACK 'N' SNACK
plate adjusts to mouth level

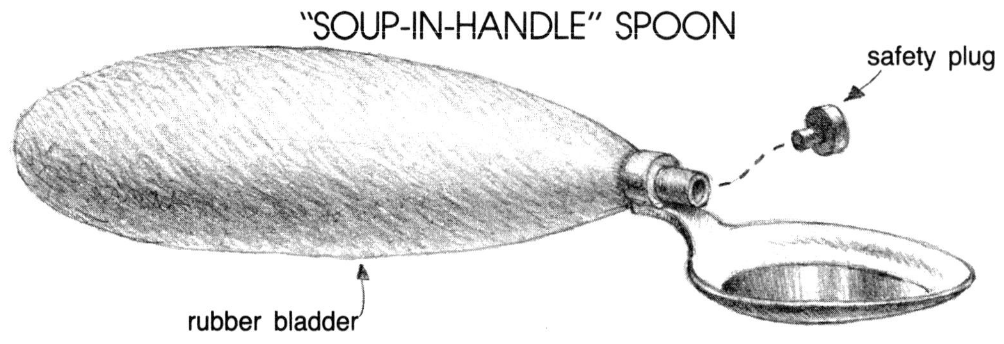
"SOUP-IN-HANDLE" SPOON
safety plug
rubber bladder

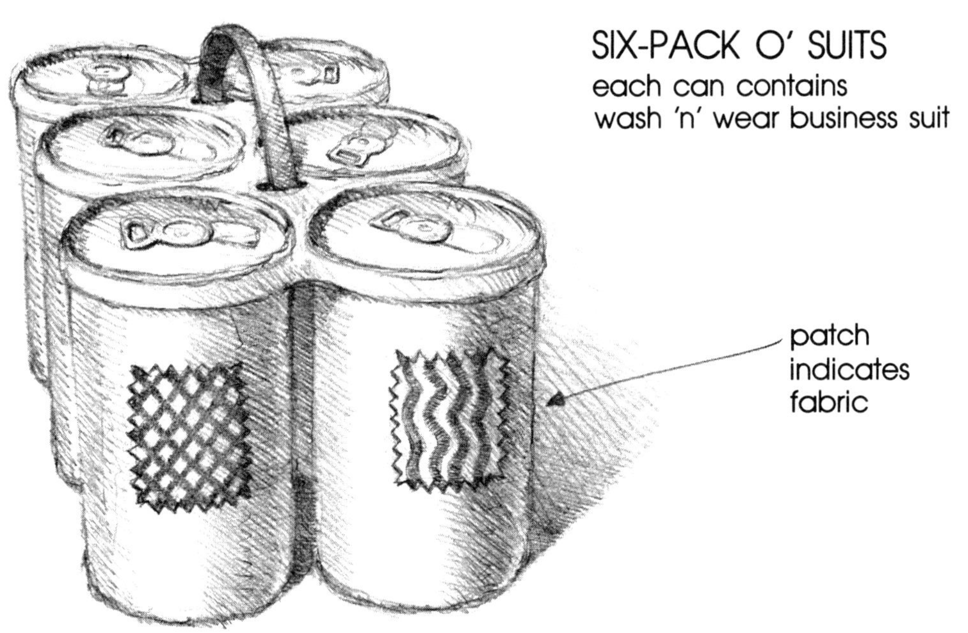
SIX-PACK O' SUITS
each can contains
wash 'n' wear business suit
patch
indicates
fabric

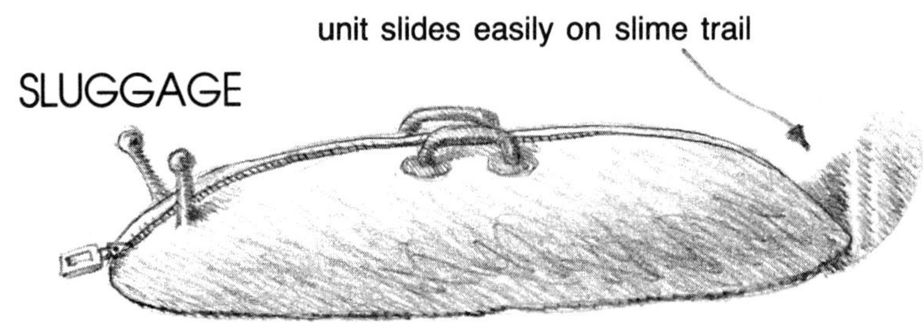
unit slides easily on slime trail
SLUGGAGE

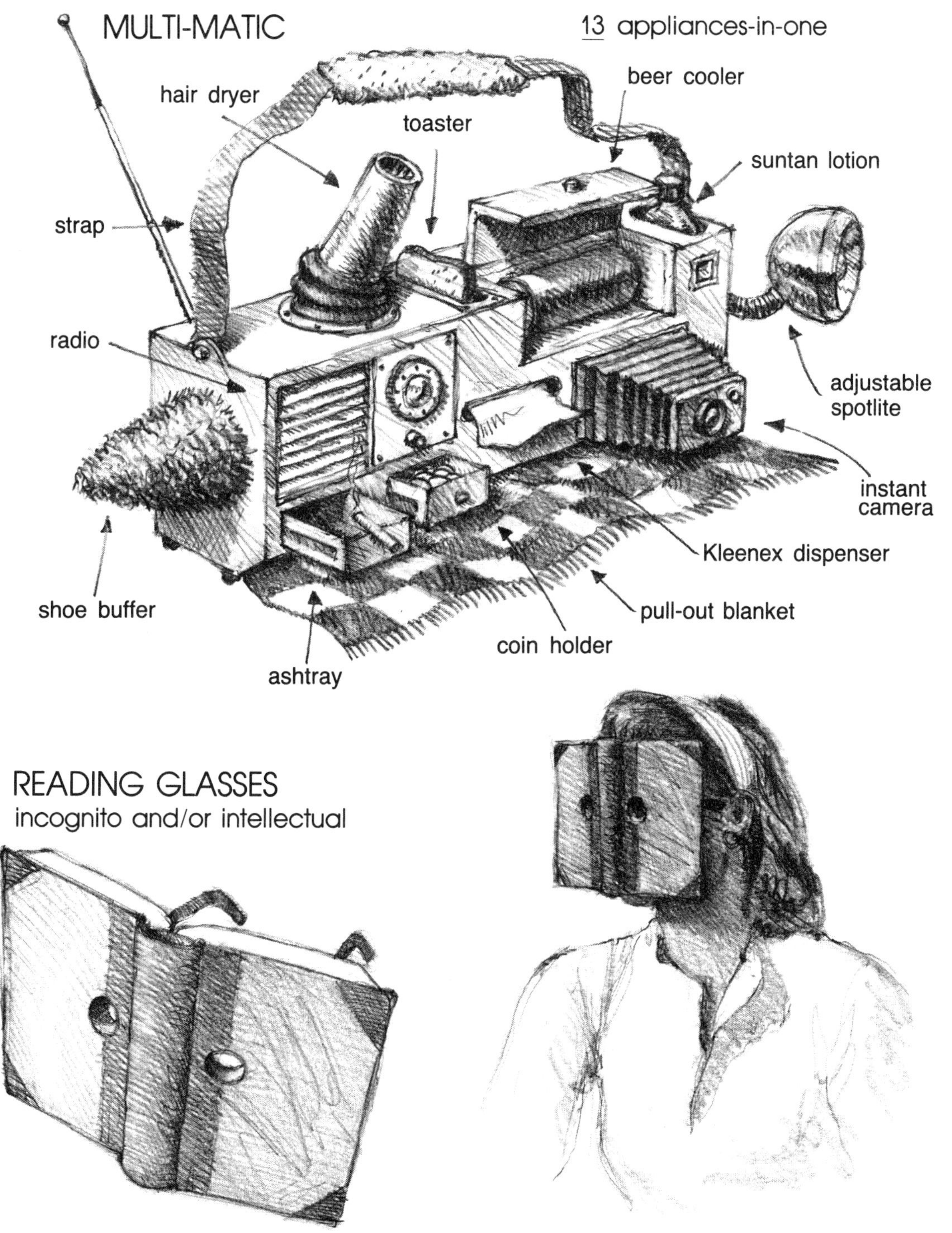
MULTI-MATIC
13 appliances-in-one
hair dryer
toaster
beer cooler
suntan lotion
strap
radio
adjustable spotlite
instant camera
Kleenex dispenser
shoe buffer
pull-out blanket
coin holder
ashtray
READING GLASSES
incognito and/or intellectual

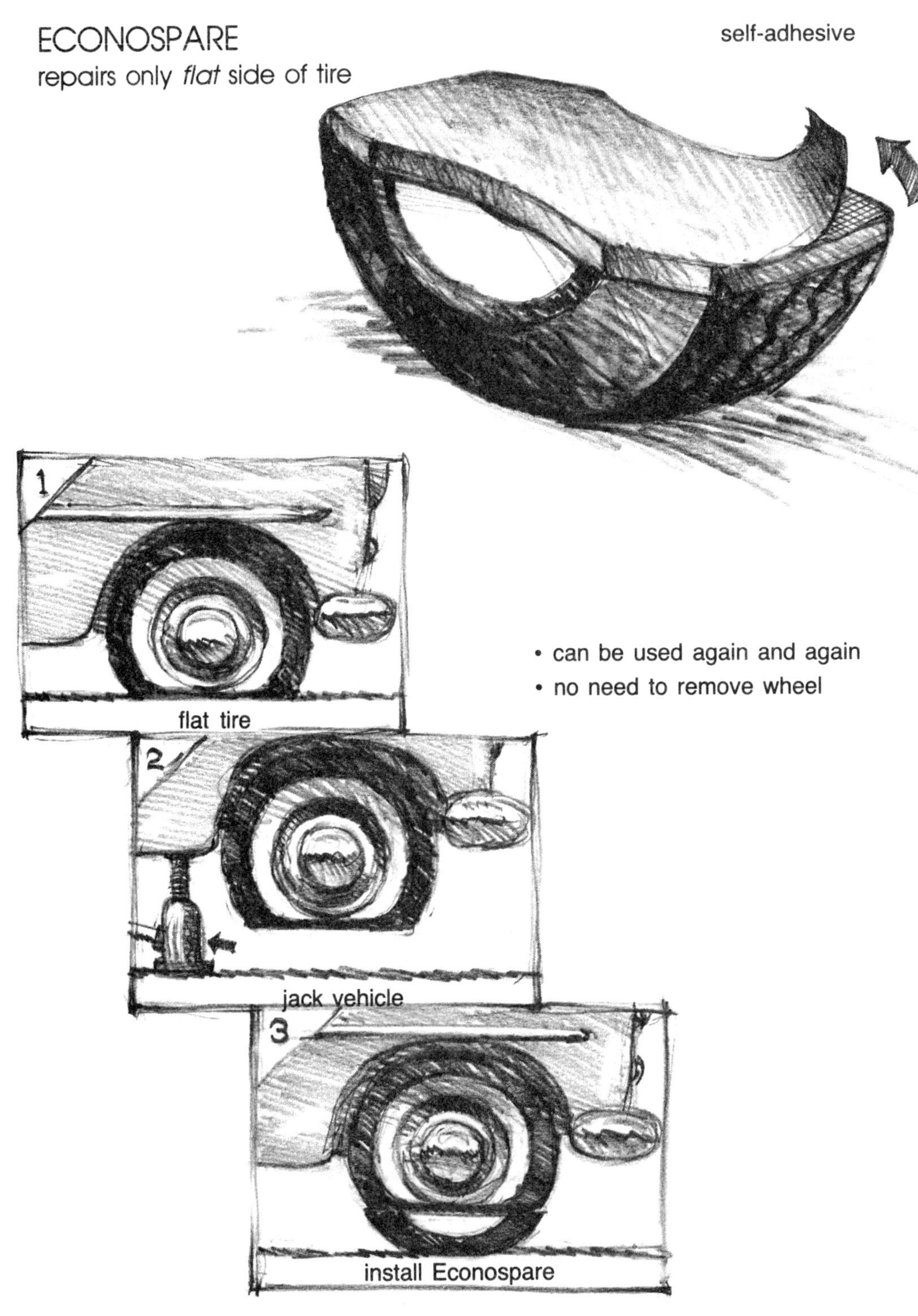

ECONOSPARE
repairs only *flat* side of tire
self-adhesive
1
flat tire
2
jack vehicle
3
install Econospare
• can be used again and again
• no need to remove wheel

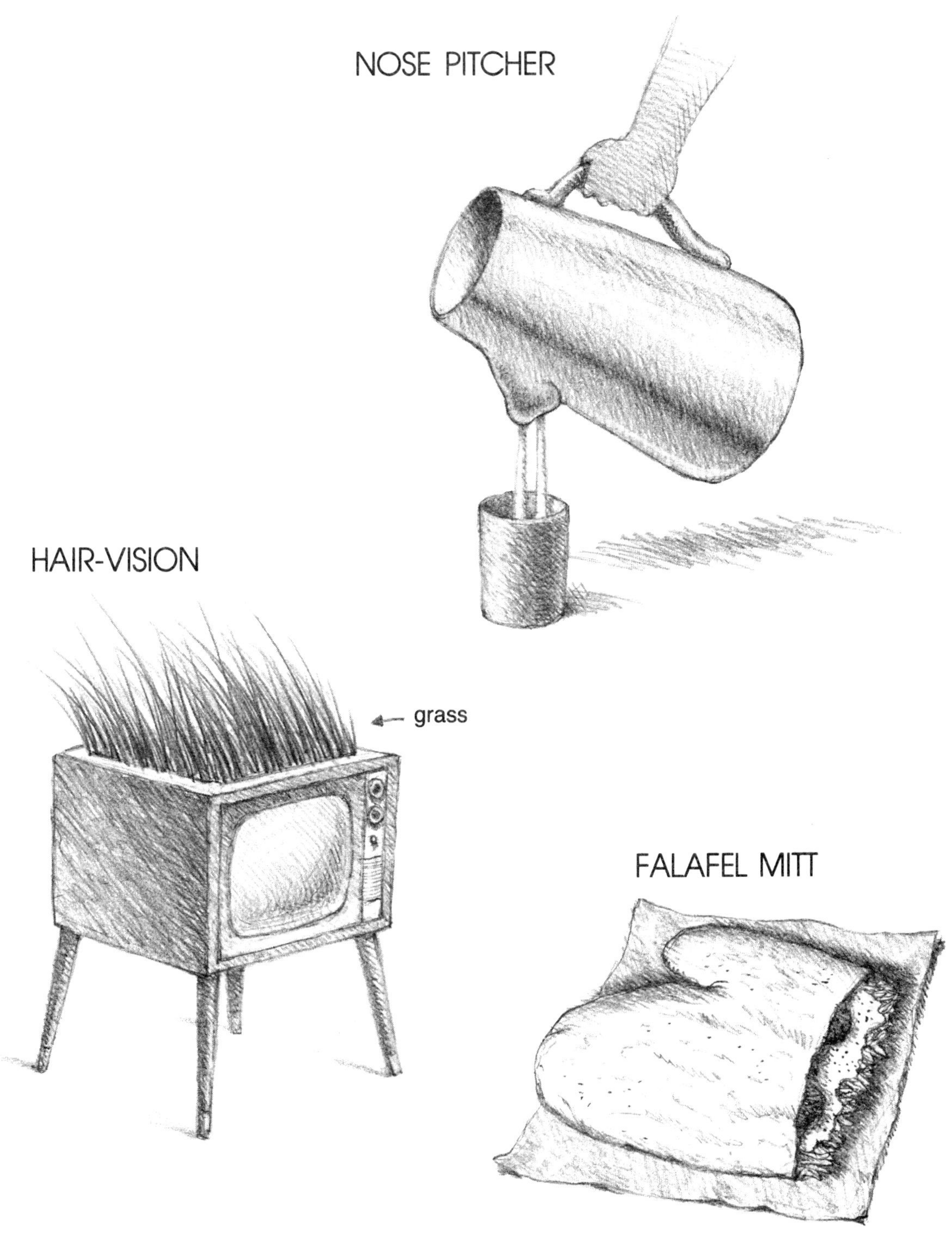
NOSE PITCHER
HAIR-VISION
grass
FALAFEL MITT

COFFEE TABLE

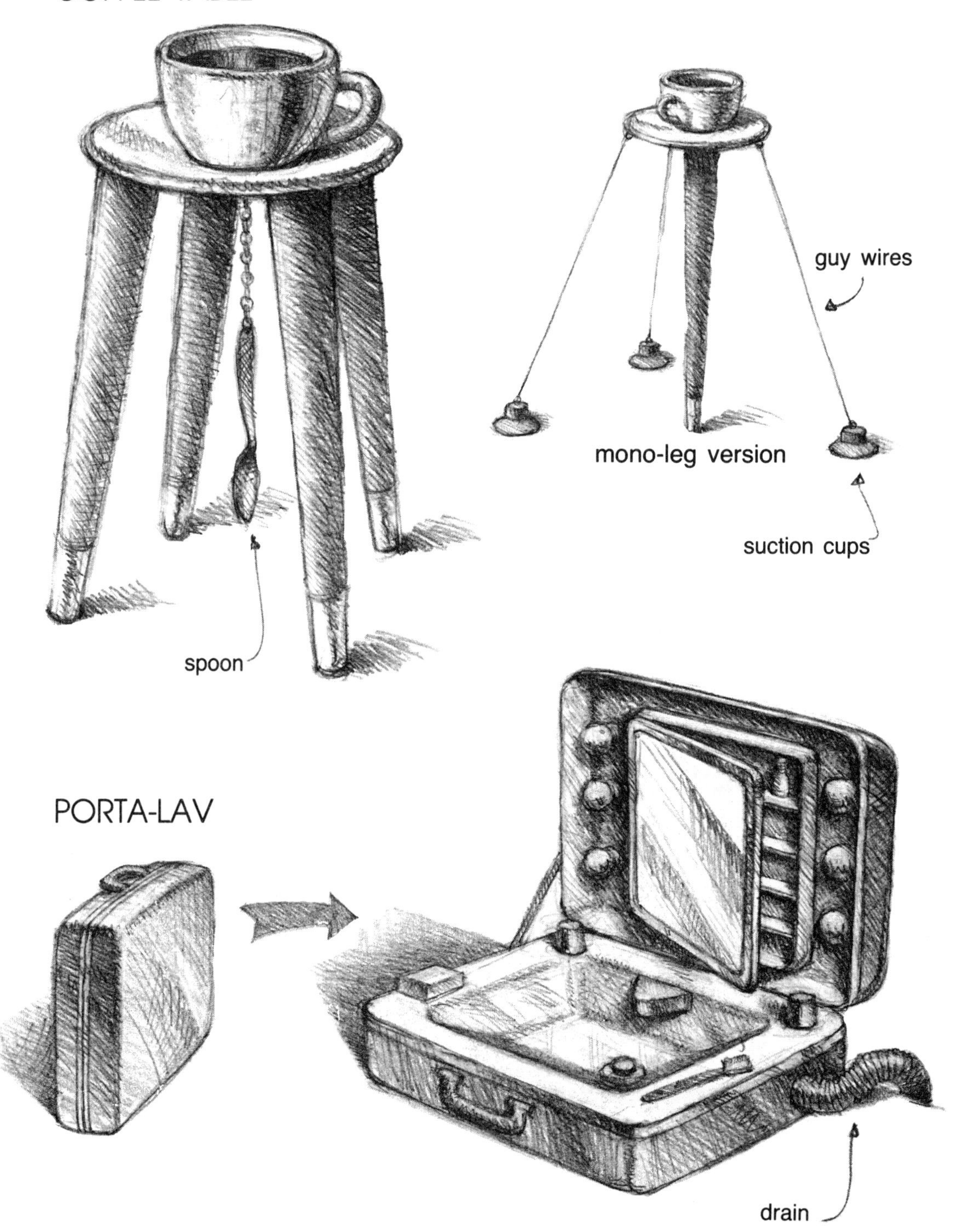

INFLATABLE "FAT LOOK" UNDERWEAR

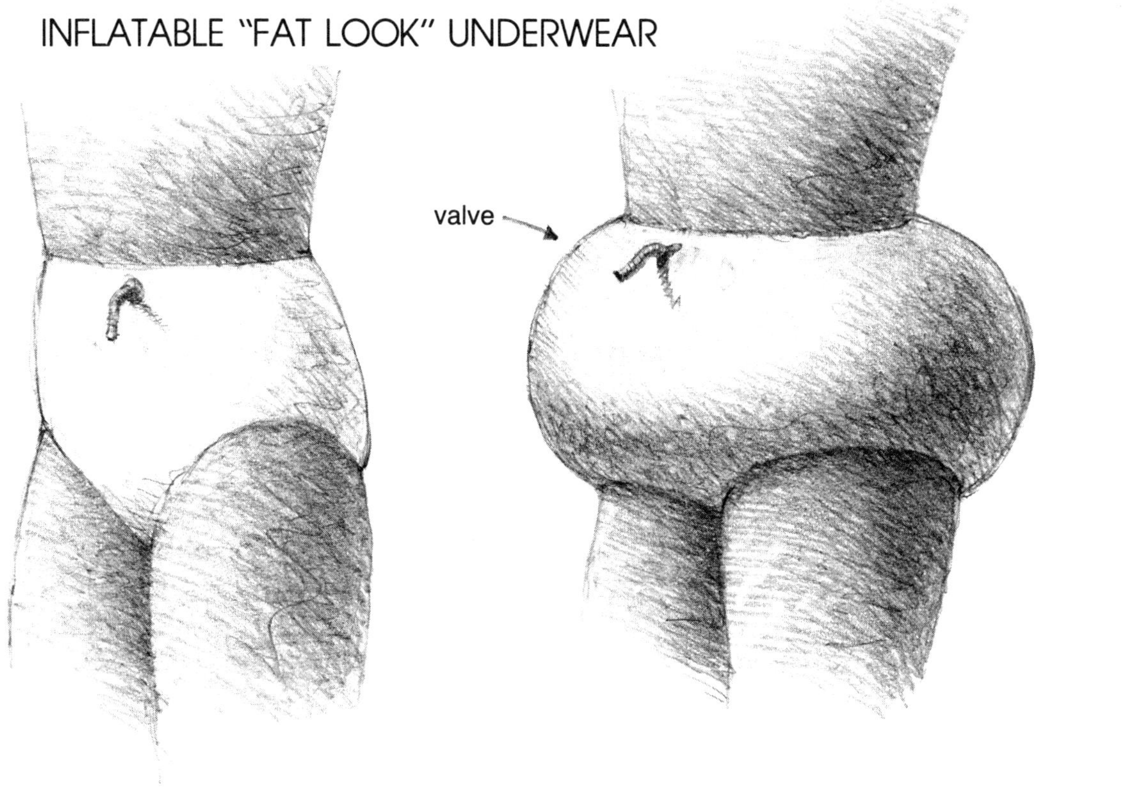

valve

HUM-DRUM

FORCE-FEEDER

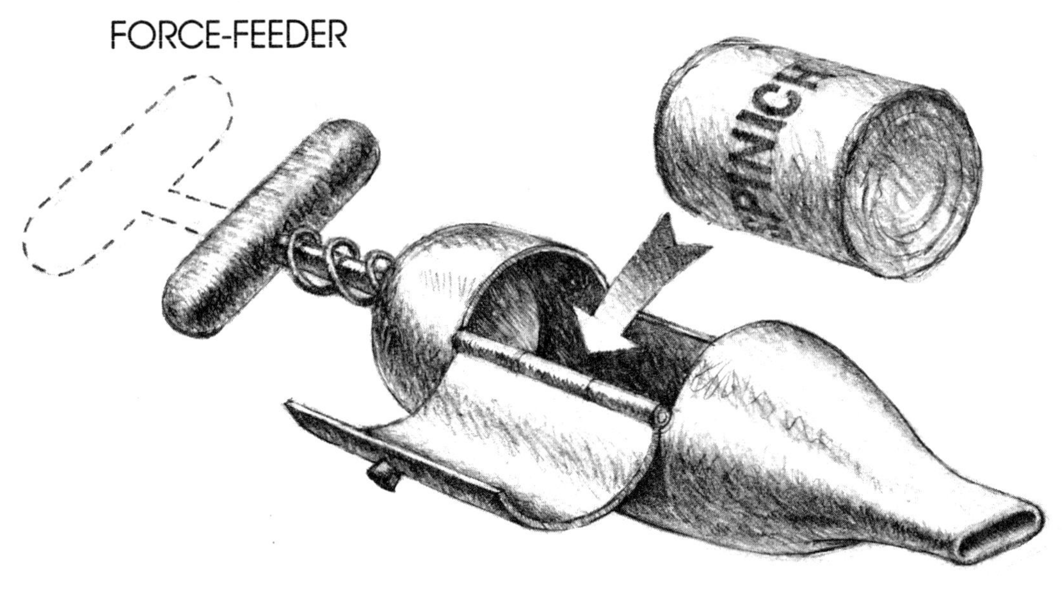

BOOK GARDEN

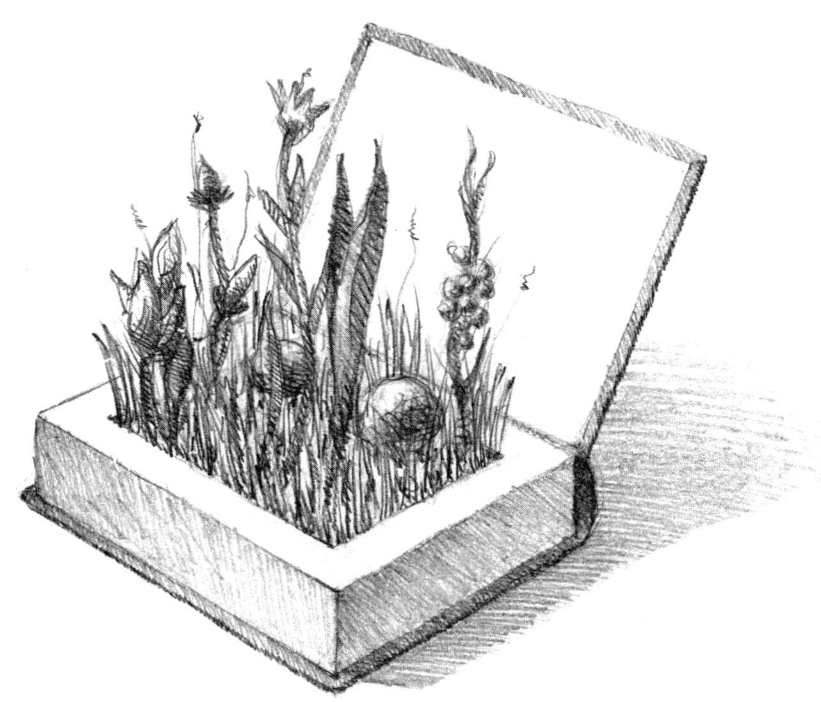

FLUSH-MOUNT EXTRA BED

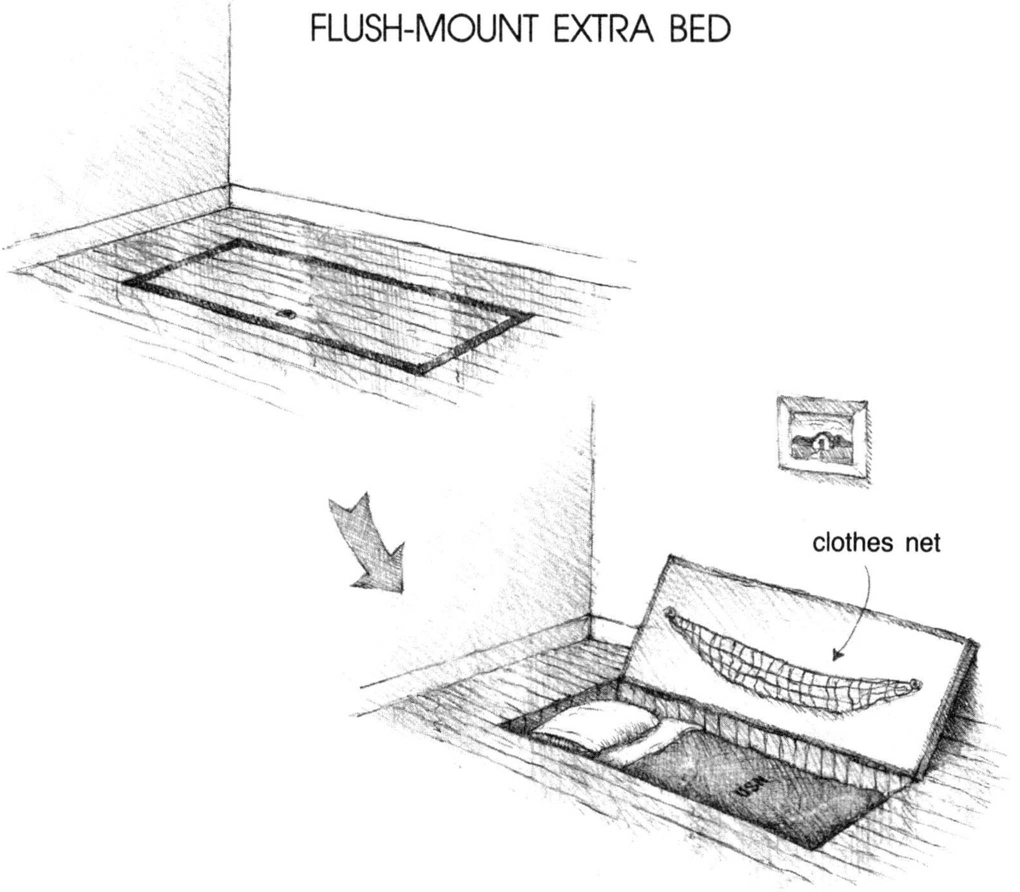

clothes net

HIDE-AWAY FURNITURE

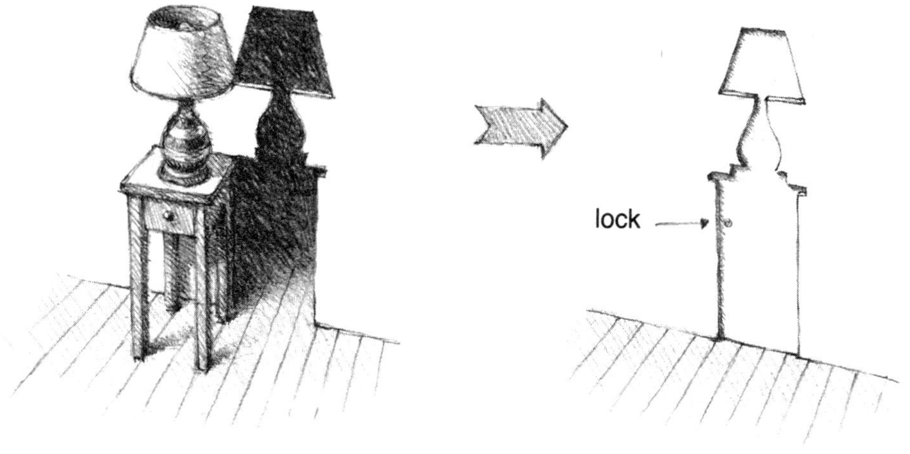

TRAV-L-TRUNK

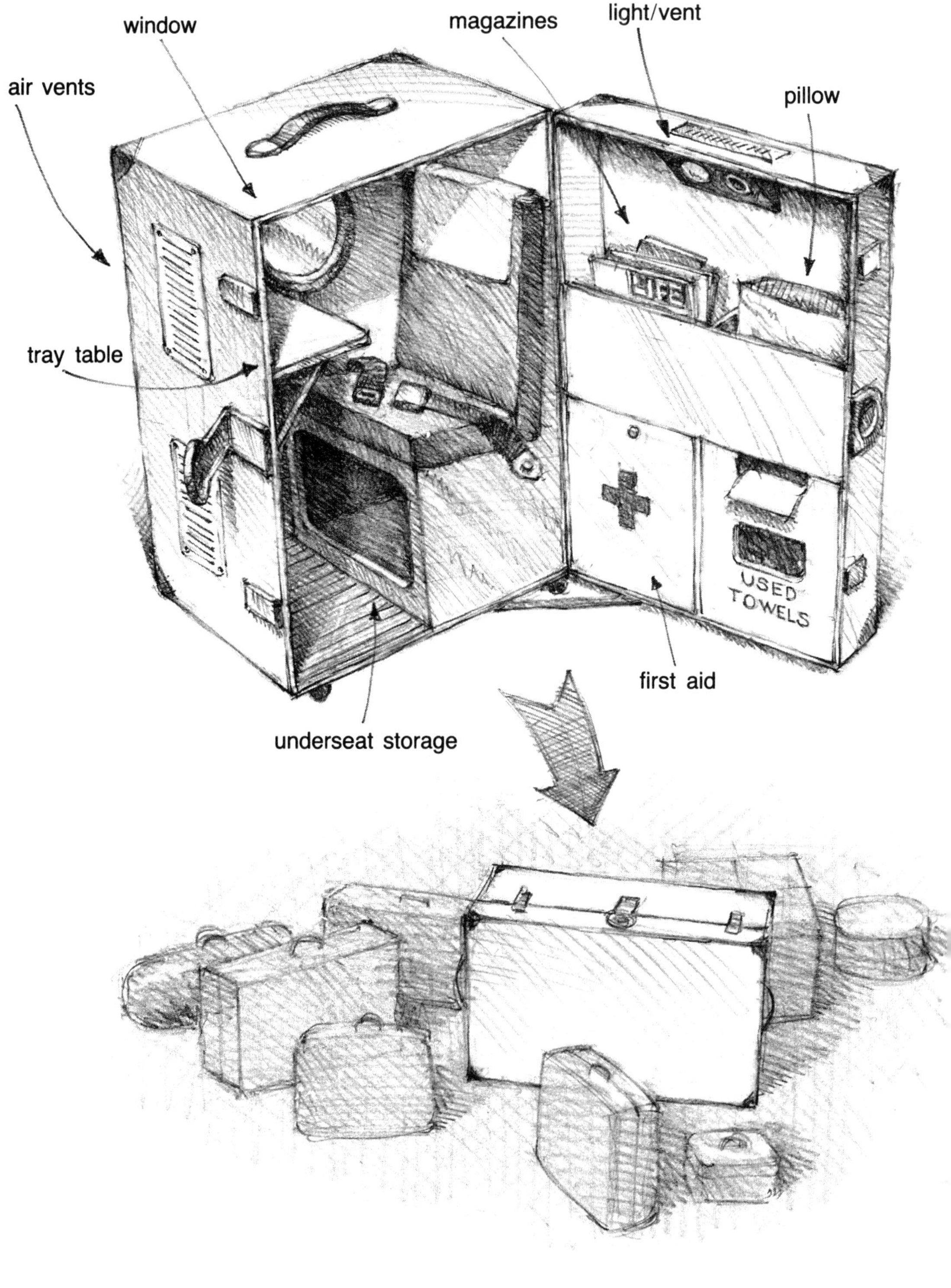

window

magazines

light/vent

air vents

pillow

tray table

USED TOWELS

first aid

underseat storage

VEIL OF FOG

UPHILL SIMULATORS

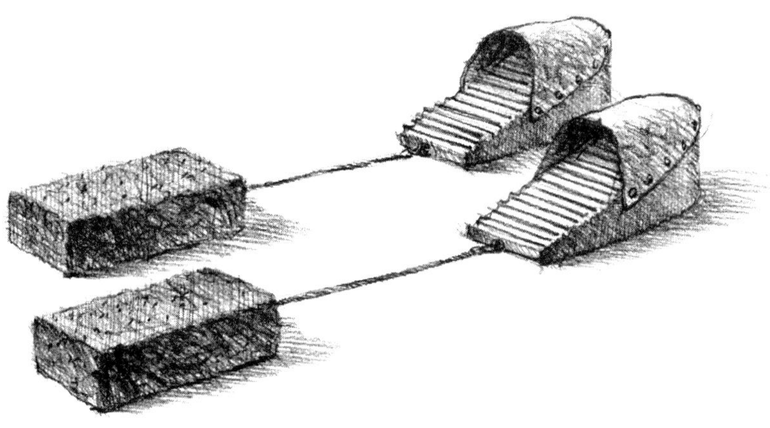

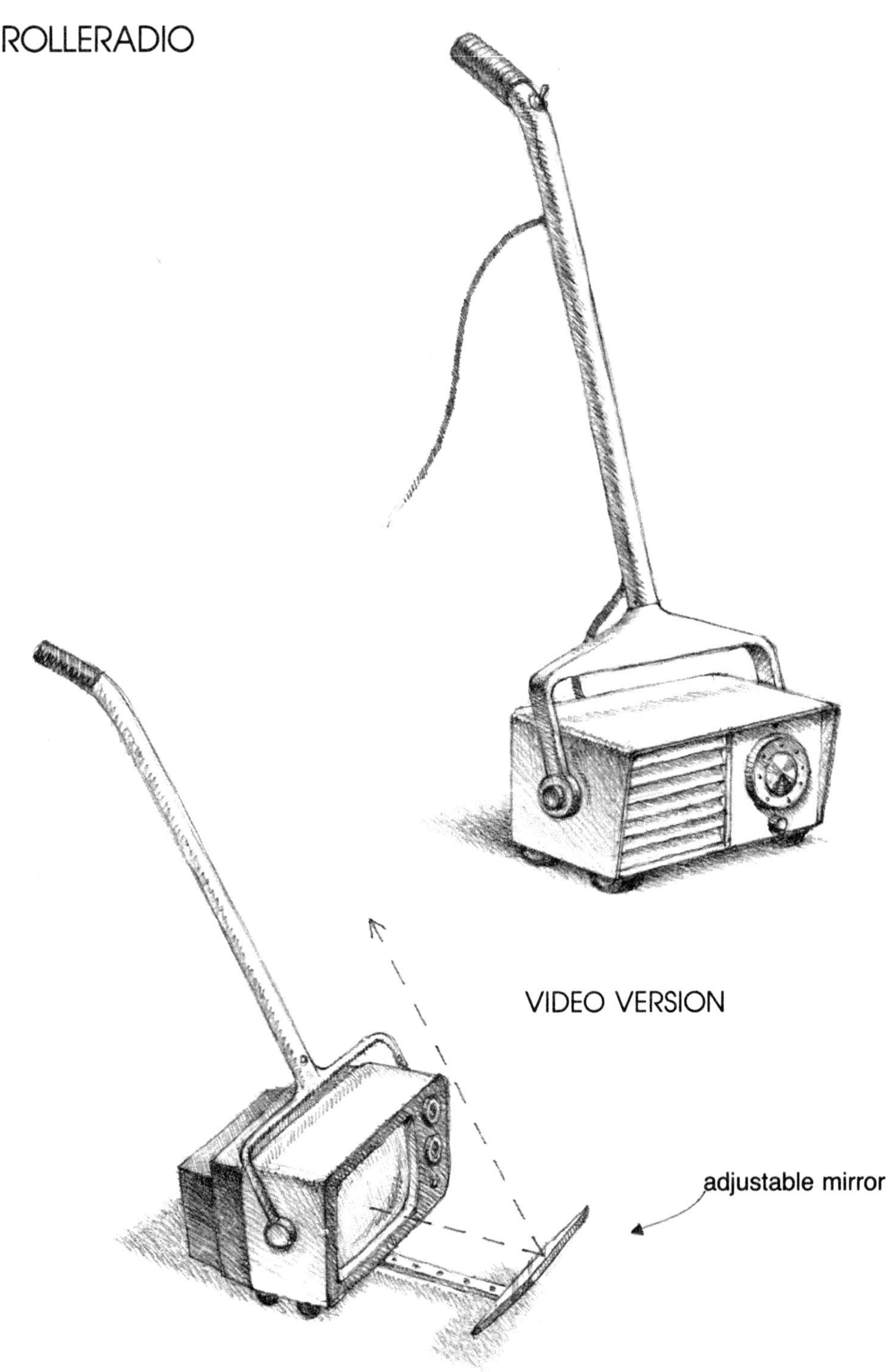
ROLLERADIO
VIDEO VERSION
adjustable mirror

MELTED LAMP

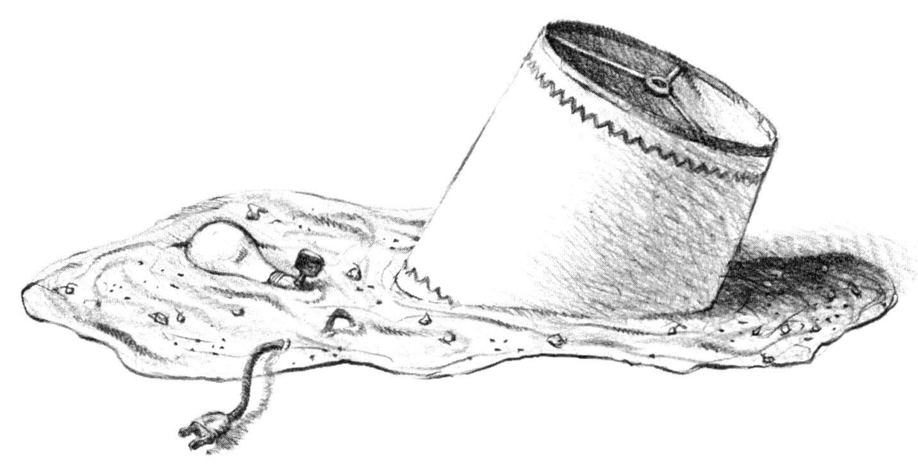

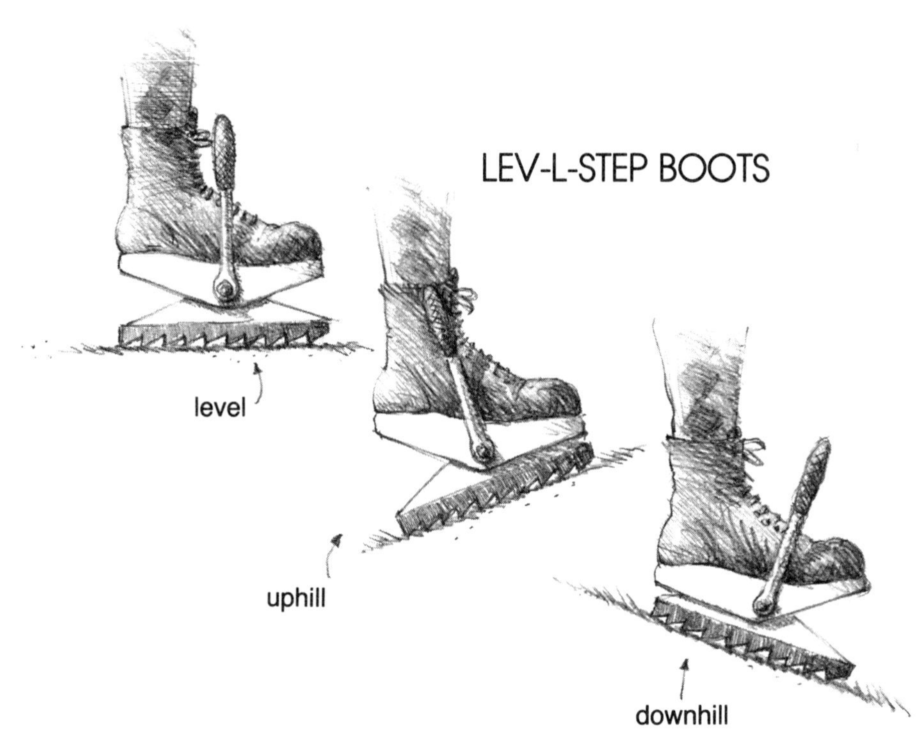
LEV-L-STEP BOOTS
level
uphill
downhill

"in use"
UNISEX RUBBING-POLE

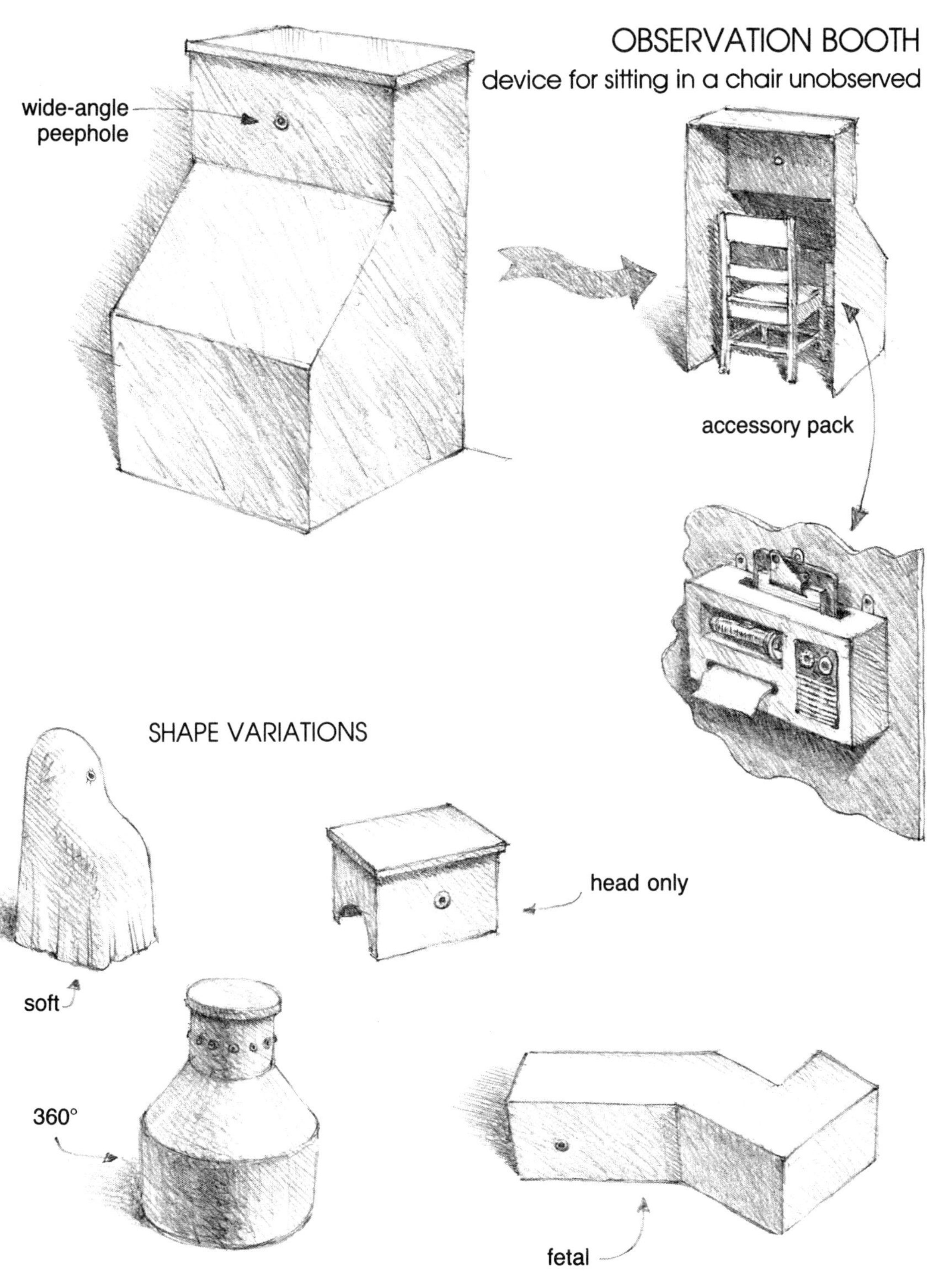

OBSERVATION BOOTH
device for sitting in a chair unobserved
wide-angle peephole
accessory pack
SHAPE VARIATIONS
head only
soft
360°
fetal

FLOOR MODEL KITCHEN TIMER

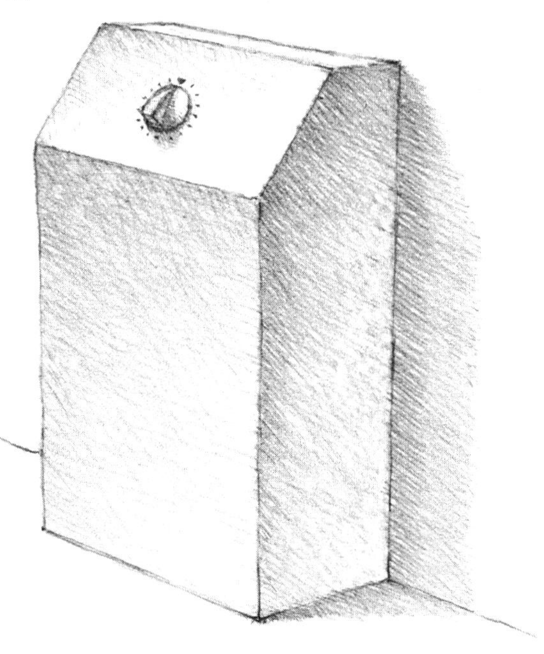

HUMAN FORK LIFT
for heavy packages

CAP INSIDE A HAT

LAWN ROWING

the ultimate test of strength

BRUSH 'N' JOT

HERO'S ENGINE
(home unit)

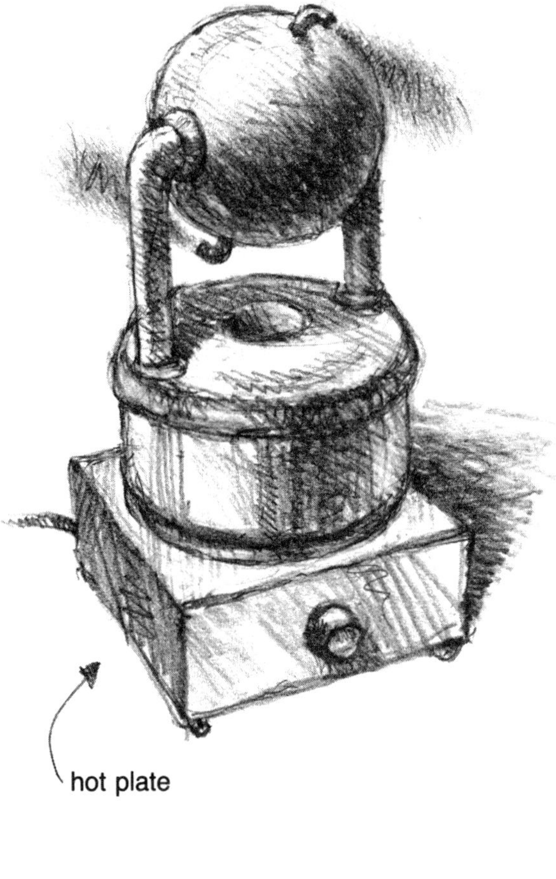

hot plate

OUTBOARD FAN

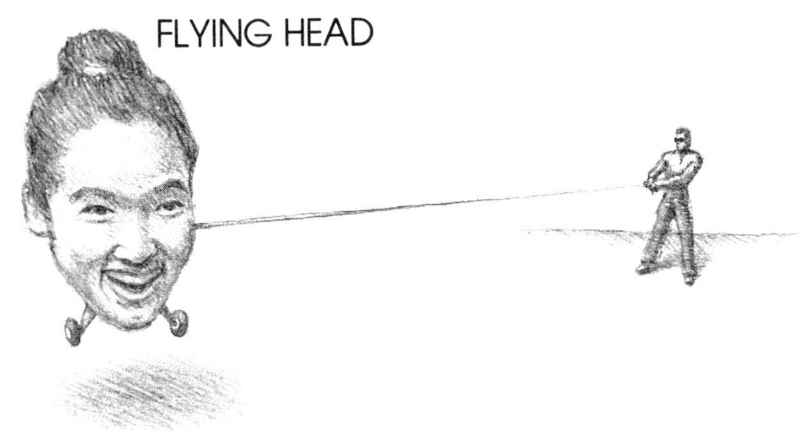

FLYING HEAD

"TRACK" COFFEE TABLE

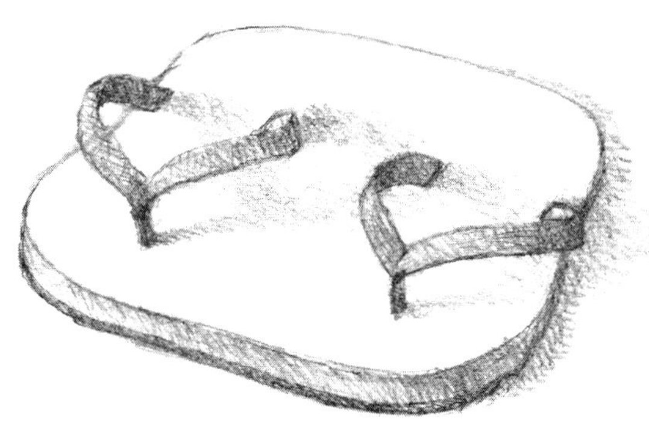

MONOTHONG

WALK-O-BIKE

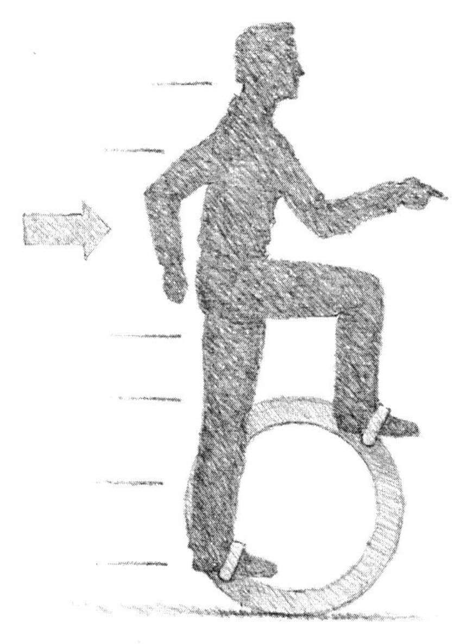

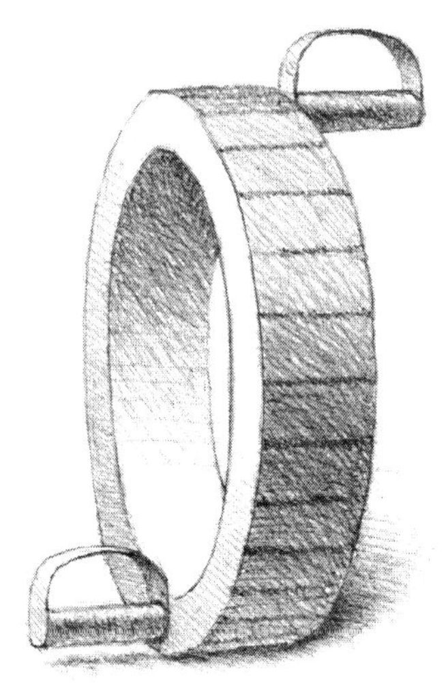

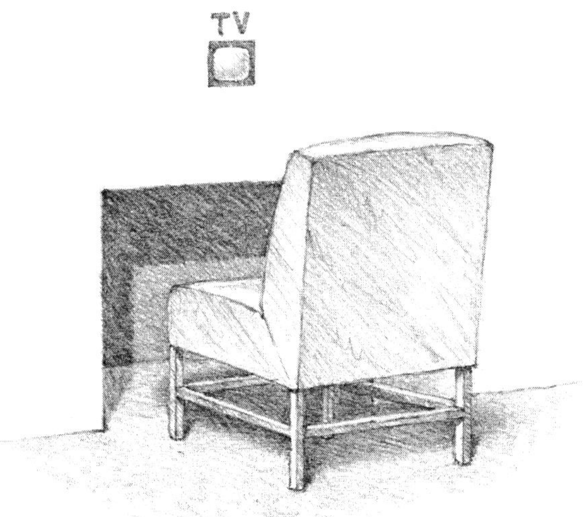

UNUSUAL LIGHT BULBS

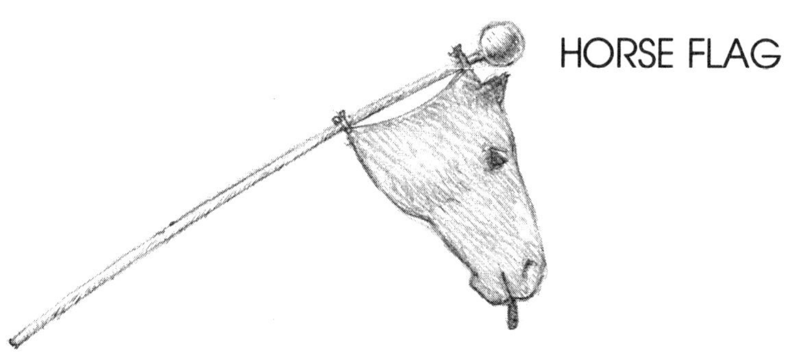
HORSE FLAG

QUONSET CAR

"THE RAPPER" WAKE-UP SYSTEM

WIG MOP

ELECTRONIC "AUDIO" FOUNTAIN

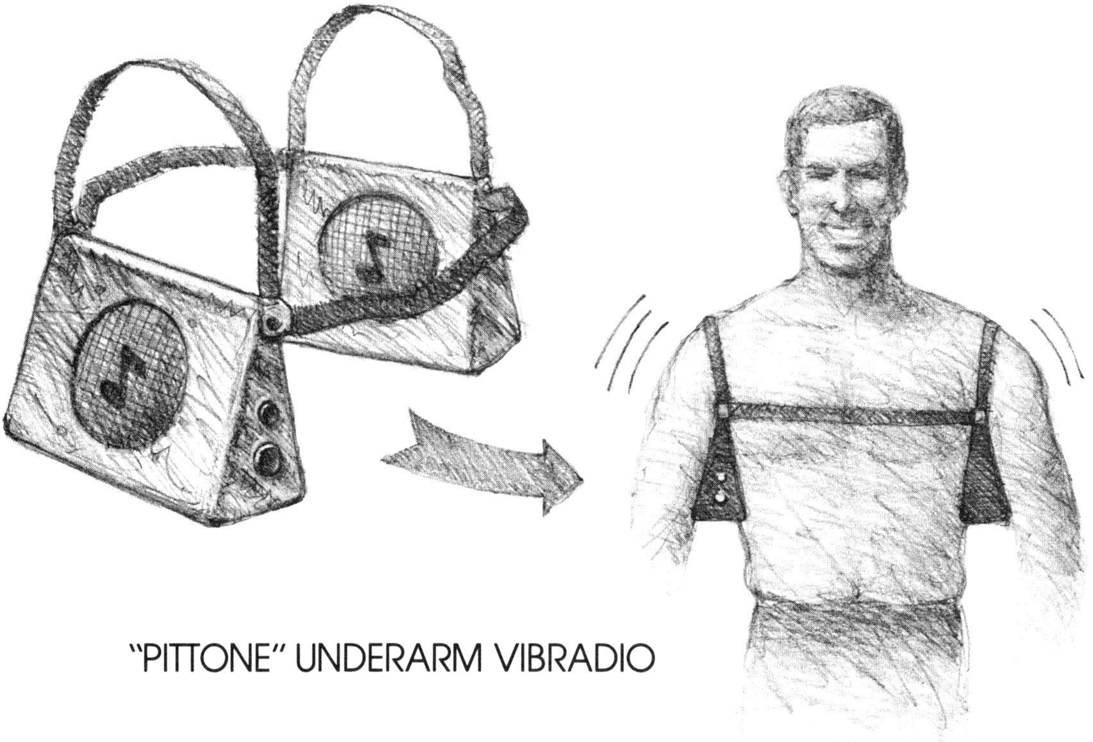

"PITTONE" UNDERARM VIBRADIO

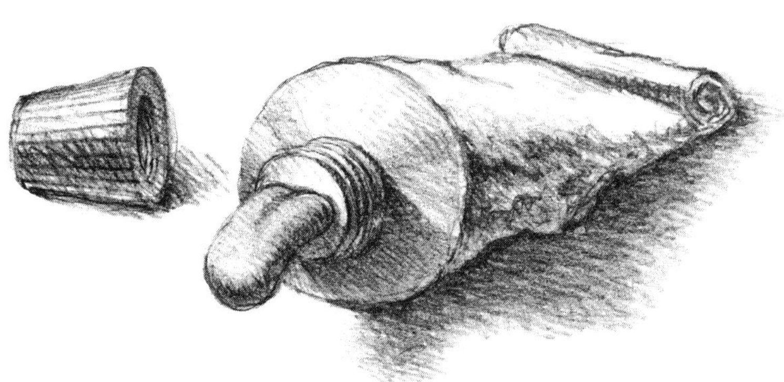

TONGUE IN A TUBE

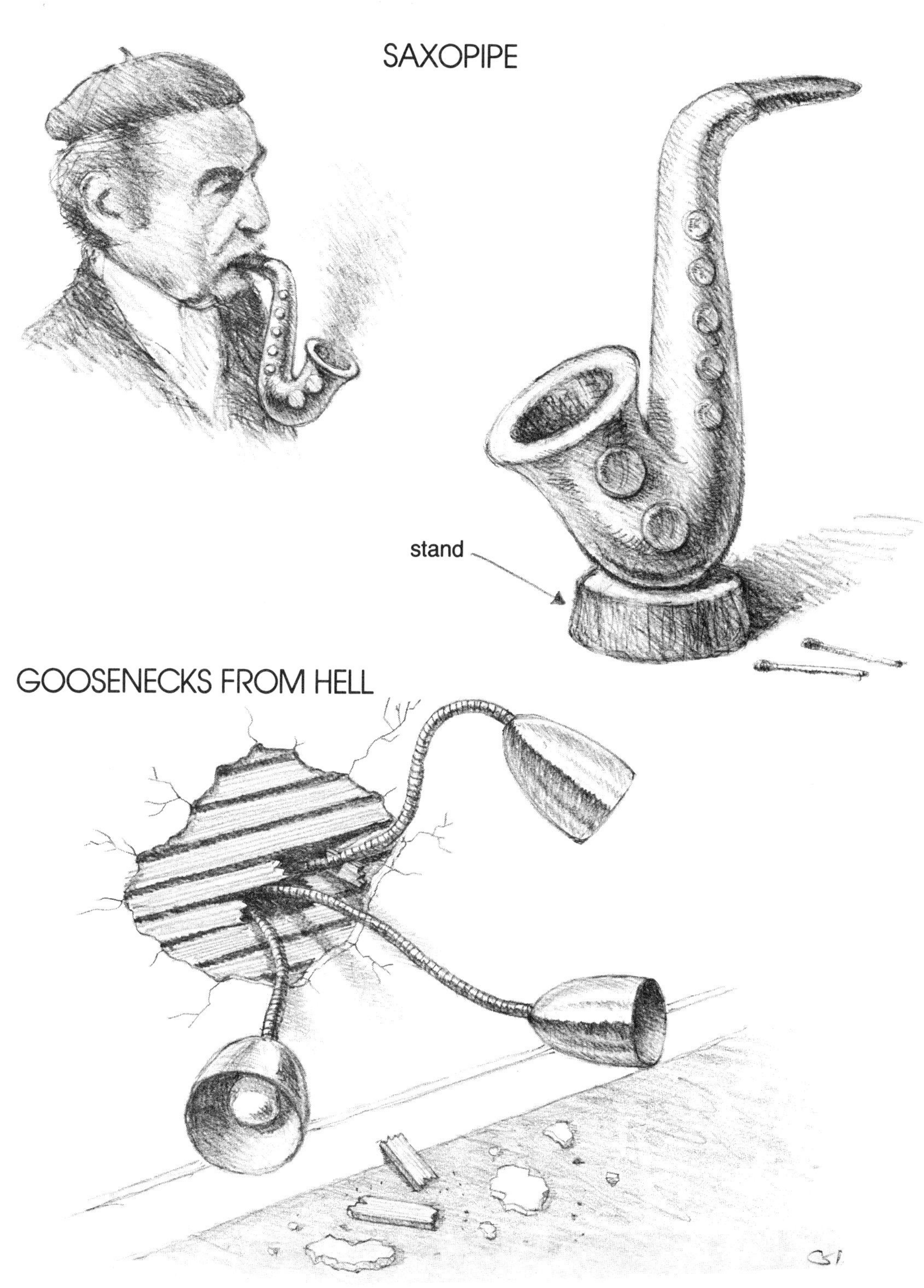
SAXOPIPE
stand
GOOSENECKS FROM HELL

TV MASKS
for program variety

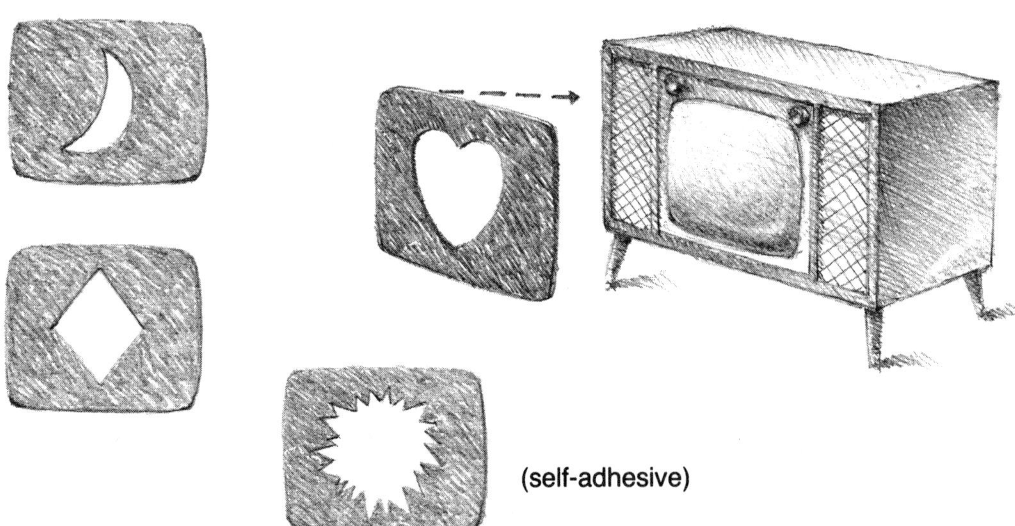

ATOMIC TIARA

"SQ' HAIR"
rectilinear coiffure

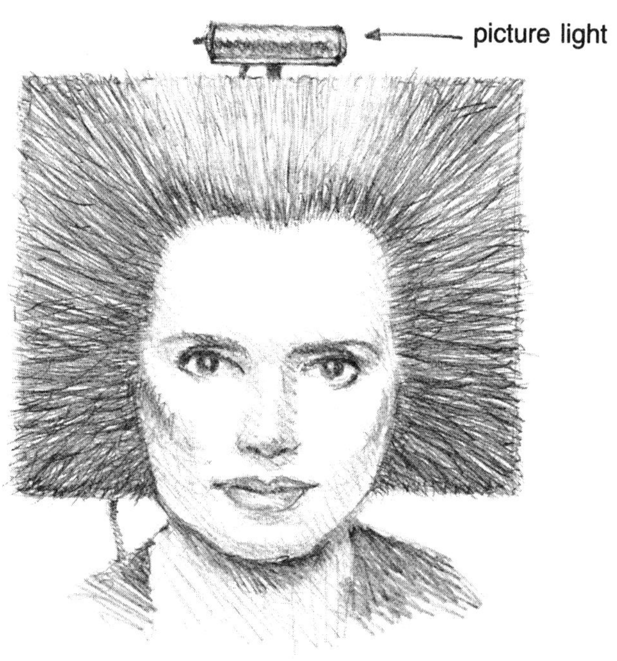

SHOE TRAILER

HORSCYCLE

TEA SNAIL
LEAKY SOFA

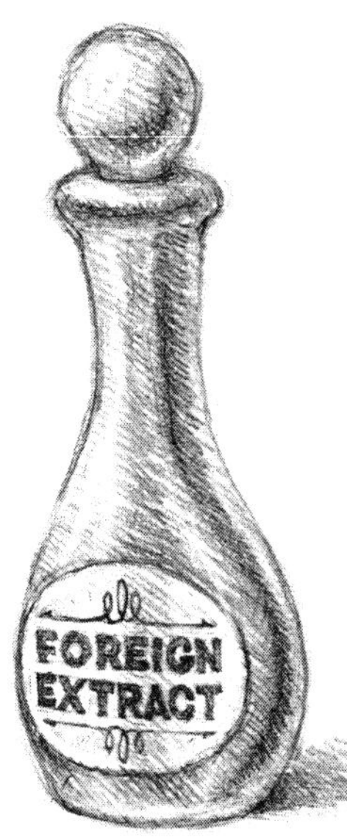
FOREIGN
EXTRACT

Sense
of
Humor

QUALMS
AND
FIGMENTS

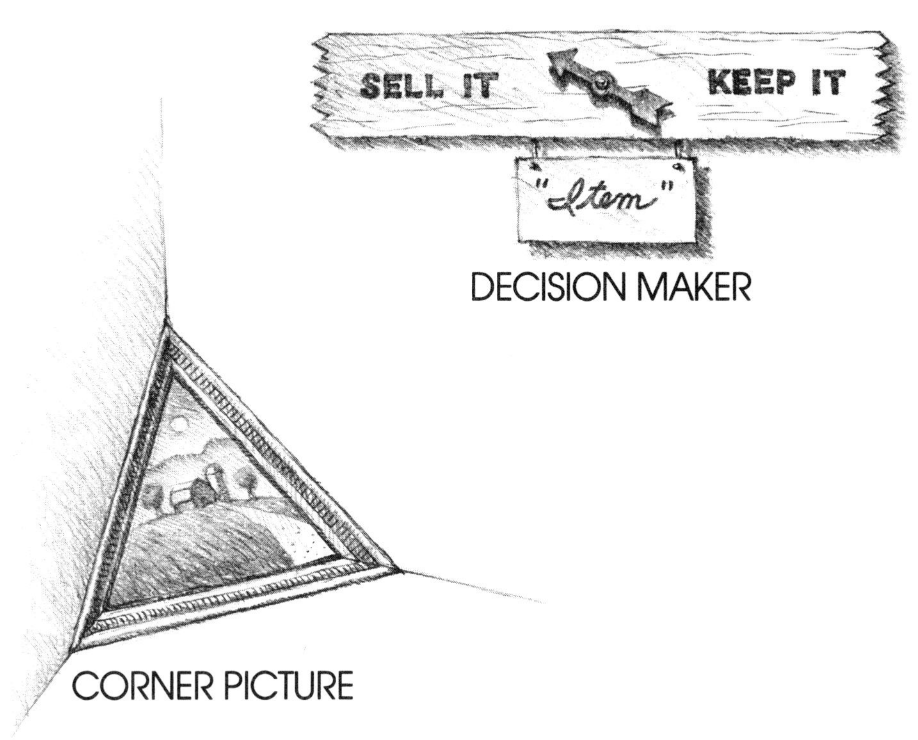
SELL IT
KEEP IT
"Item"
DECISION MAKER
CORNER PICTURE

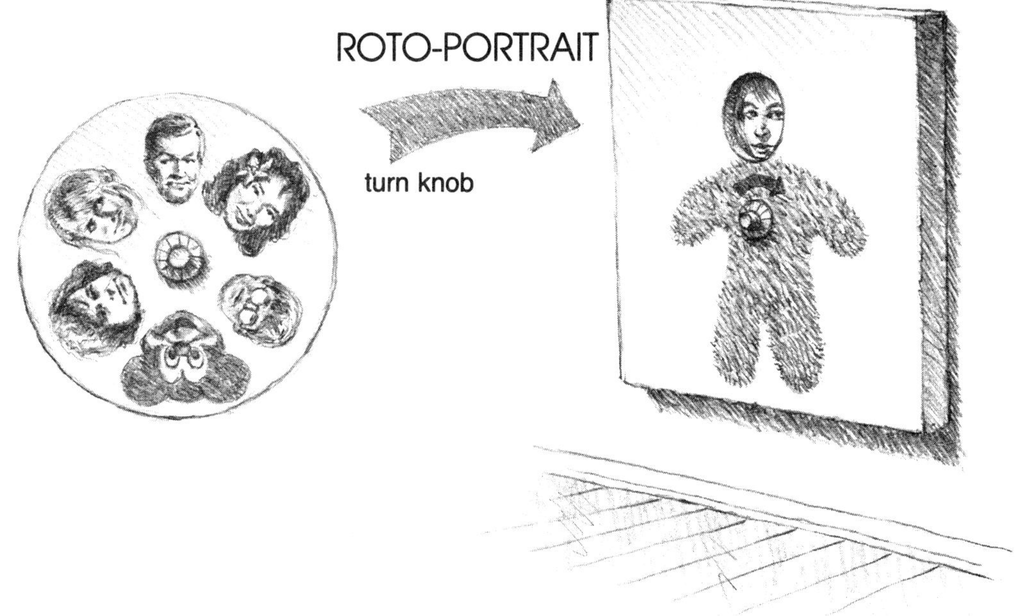
ROTO-PORTRAIT
turn knob

SEX-SPEX

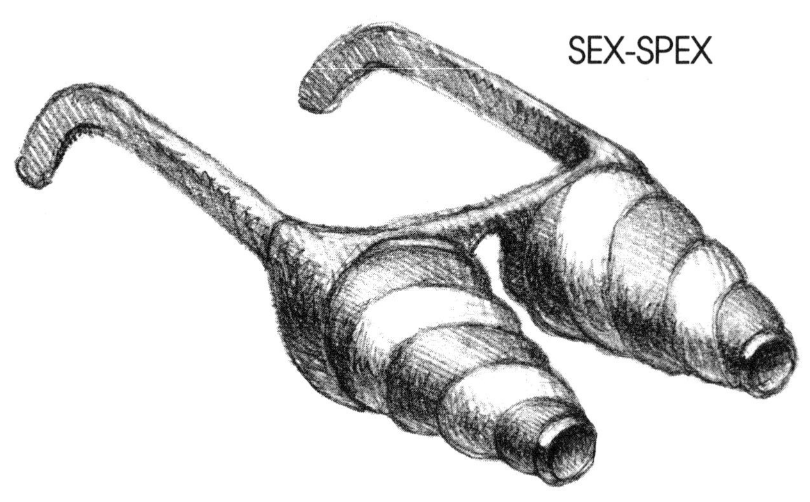

GENDEROMETER

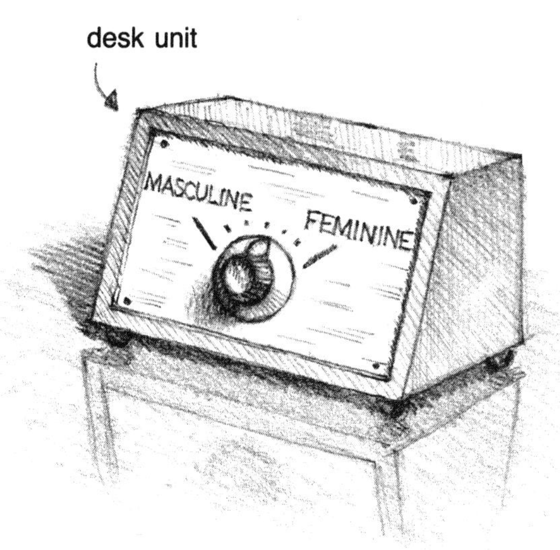

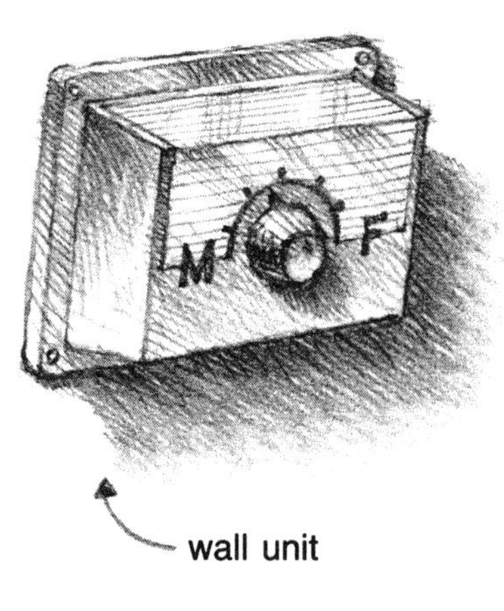

POP-UP SATURN

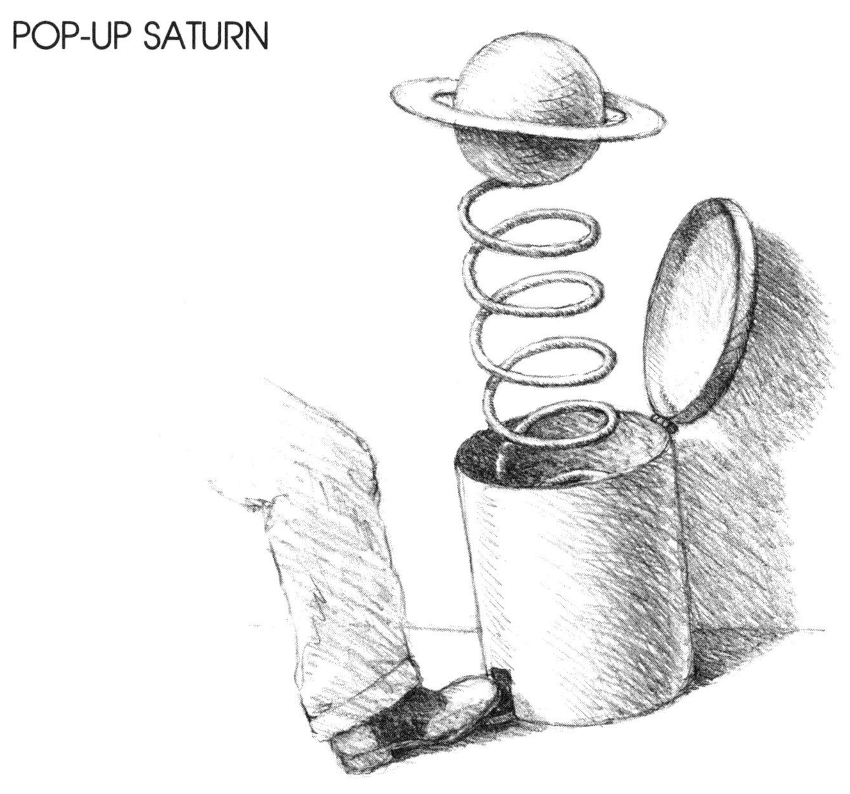

STEREO-TYPE

SPIT TUNE

GRAMMAR INSTRUCTION T-SHIRTS

"NEVER-LOSE" HAT

"TANKER" BEVERAGE SYSTEM

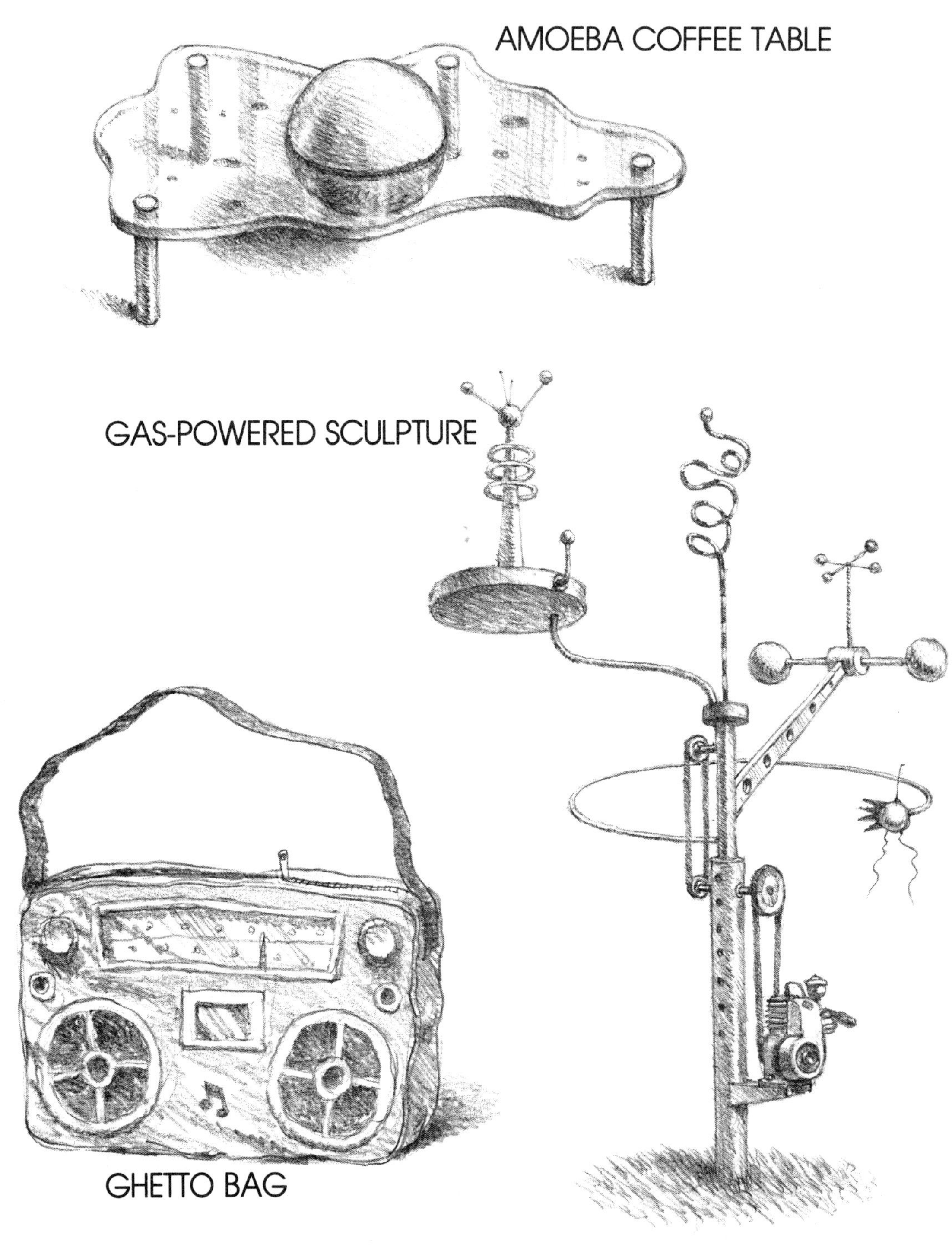
AMOEBA COFFEE TABLE
GAS-POWERED SCULPTURE
GHETTO BAG

BACKWARDS STAIRS

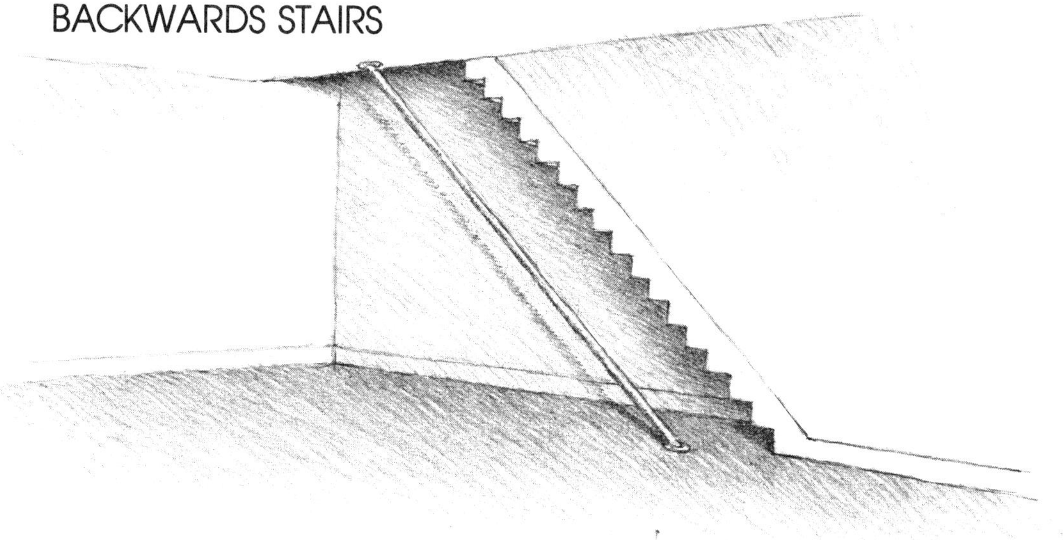

SEDAN CHAIR

WORLD WAFFLE IRON

360° ROCKER

sofa version

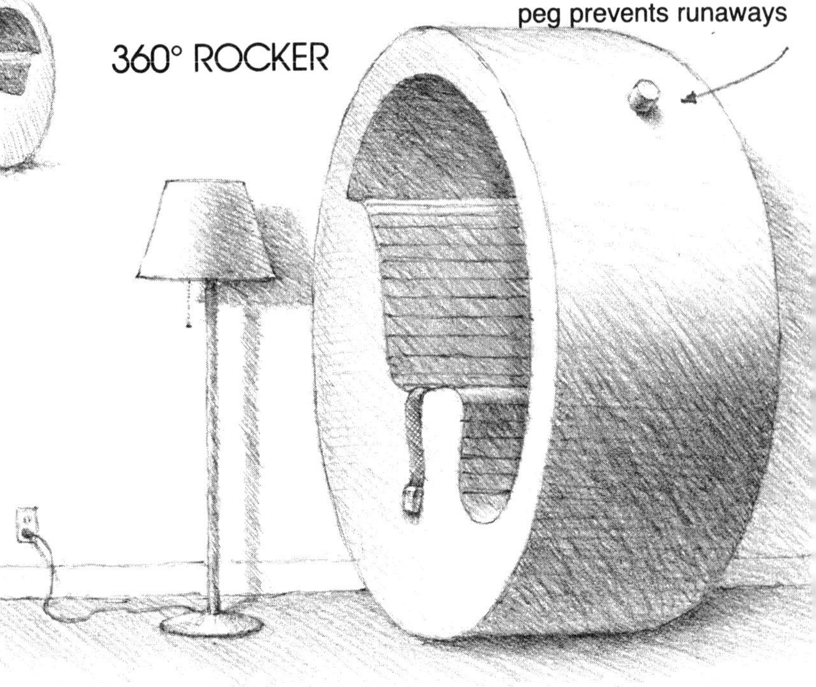

peg prevents runaways

ZIP-SOLE SHOES

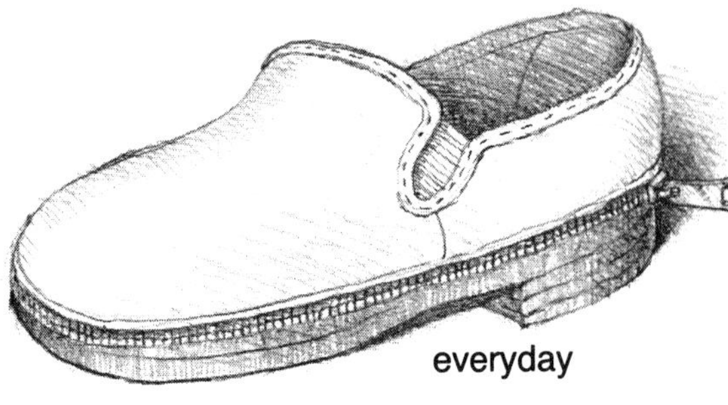

everyday

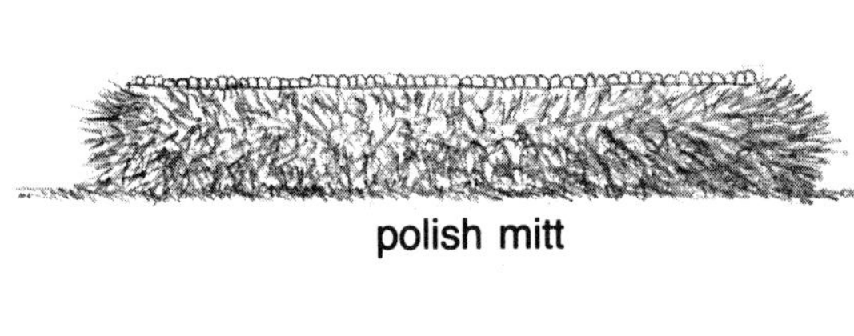

polish mitt

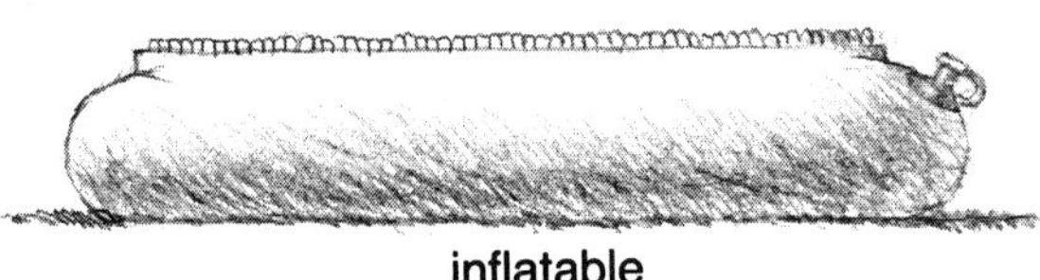

inflatable

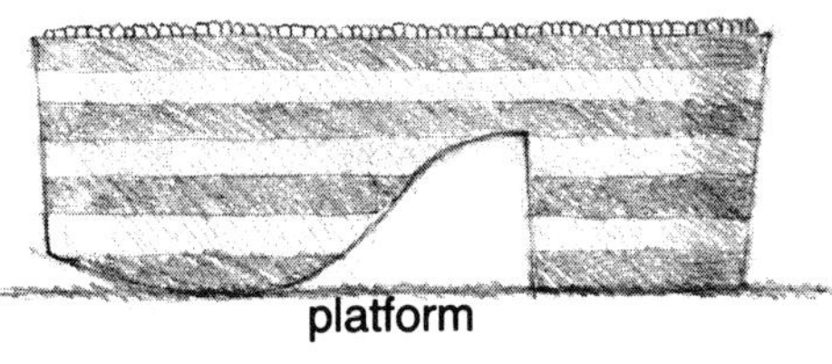

platform

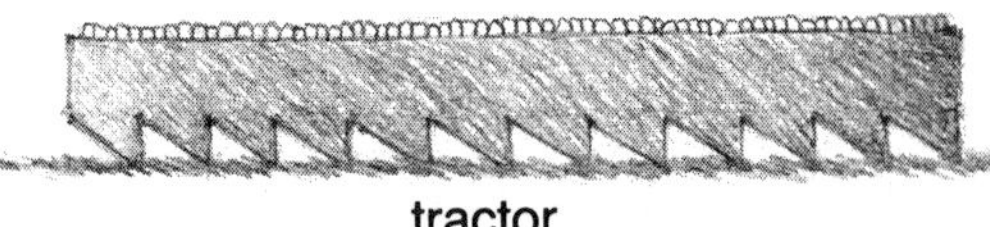

tractor

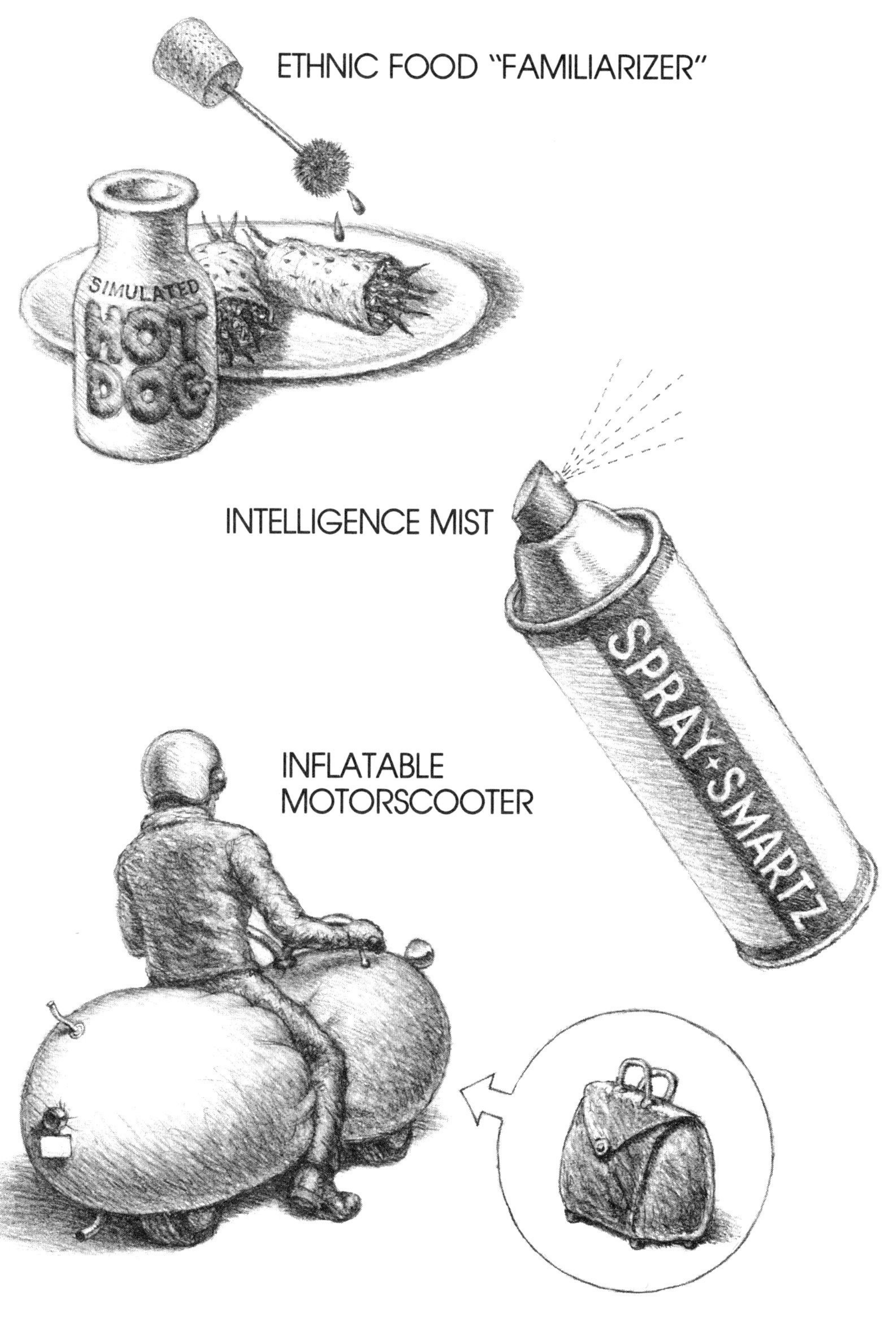
ETHNIC FOOD "FAMILIARIZER"
SIMULATED
HOT DOG
INTELLIGENCE MIST
SPRAY★SMARTZ
INFLATABLE
MOTORSCOOTER

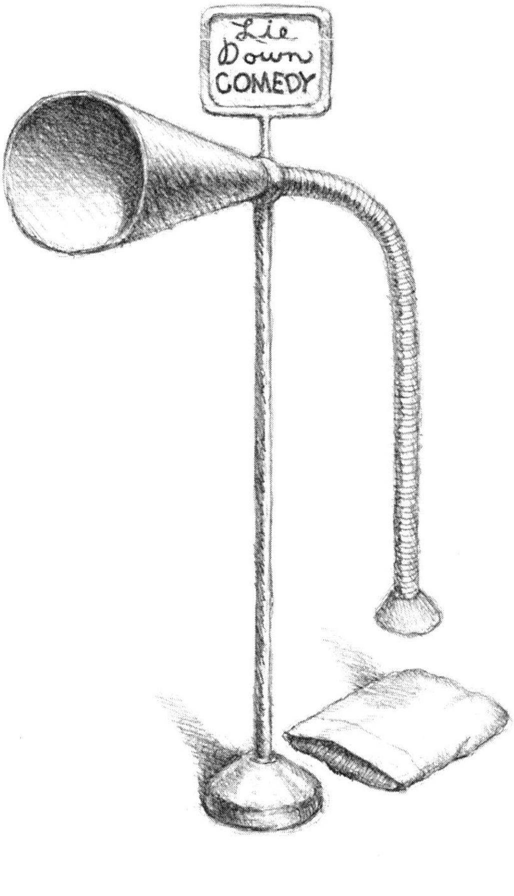
Lie Down COMEDY

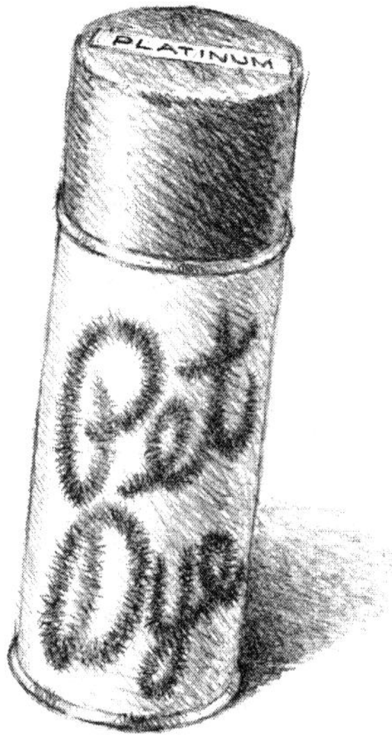
PLATINUM
Pet Dye

shrouds prevent pedestrian collisions

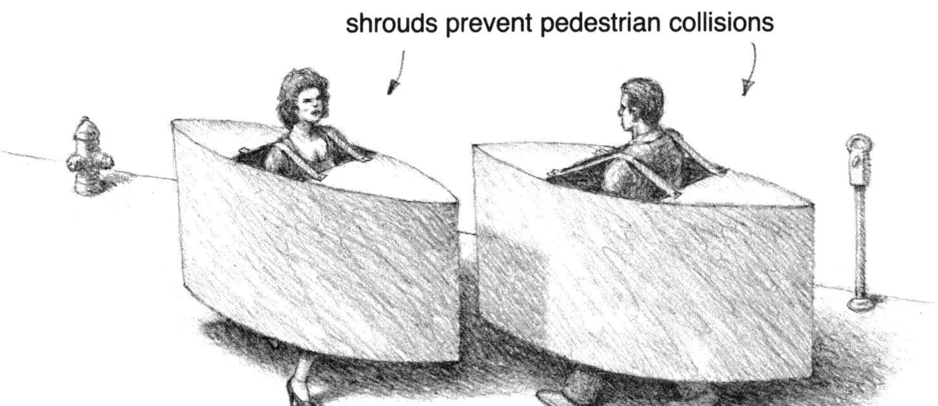

JAVA-NILE

HUMAN WASHING MACHINE
agitator
helmet

TOE-LAMP
shoe strap
BATTS

ONE-STROKE COMB

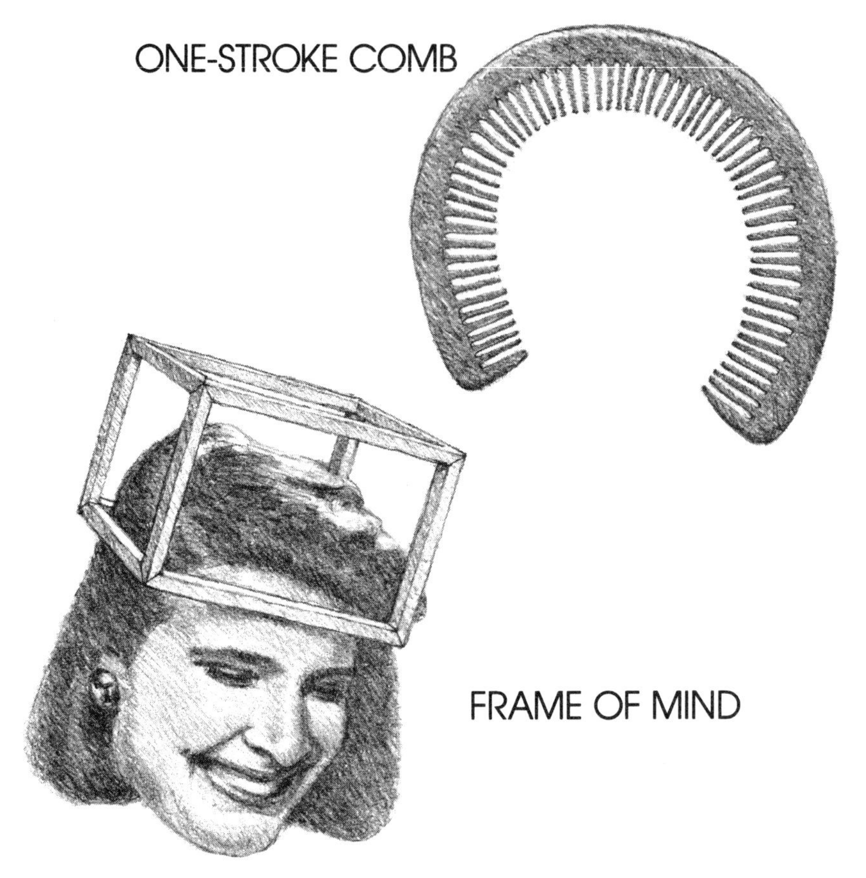

FRAME OF MIND

WINDSHIELD WIPER RAINBOW MAKER

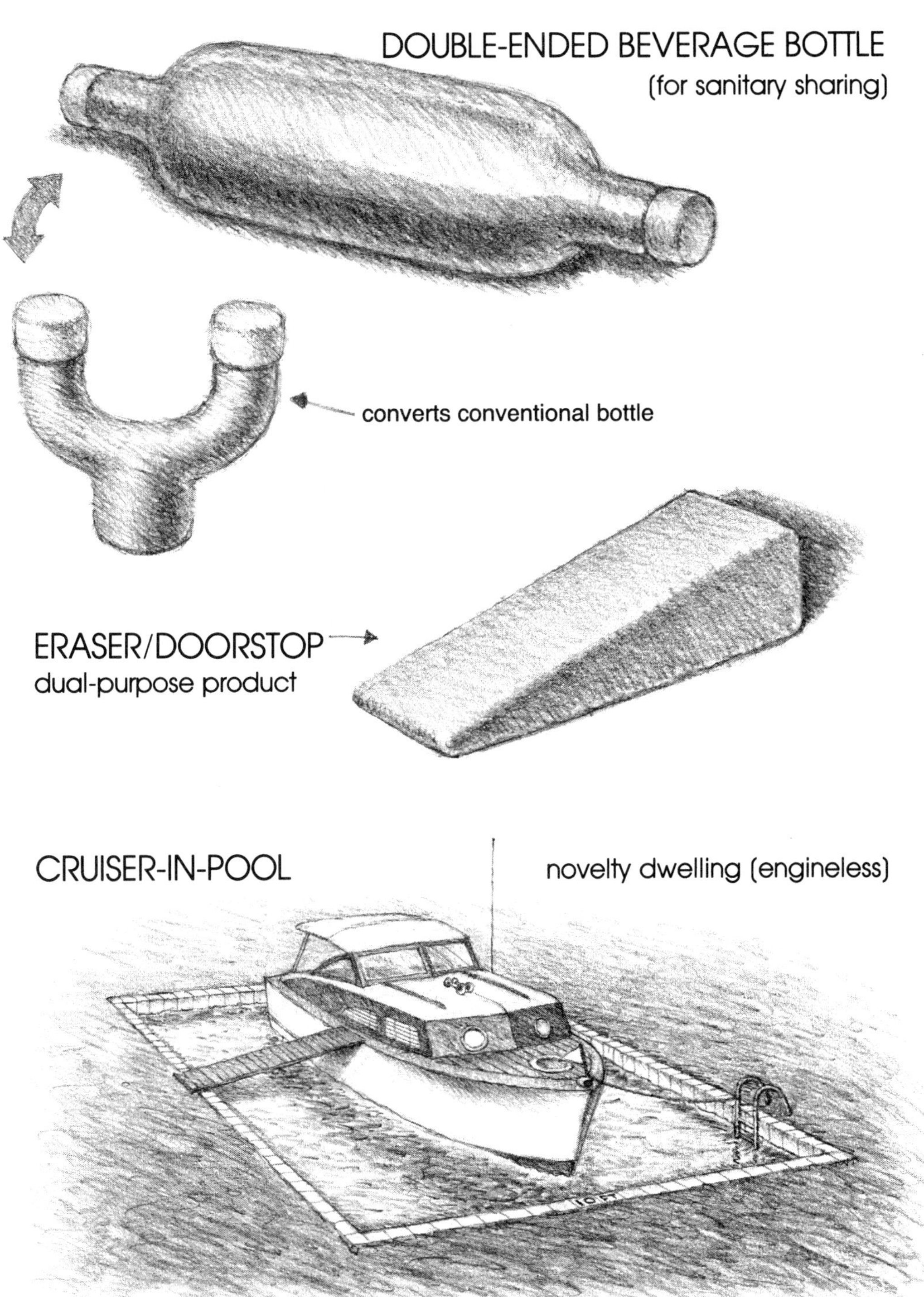
DOUBLE-ENDED BEVERAGE BOTTLE
(for sanitary sharing)
converts conventional bottle
ERASER/DOORSTOP
dual-purpose product
CRUISER-IN-POOL
novelty dwelling (engineless)
10 FT

"MAKE 'EM SMILE"
camera attachment

HOME DITCHER

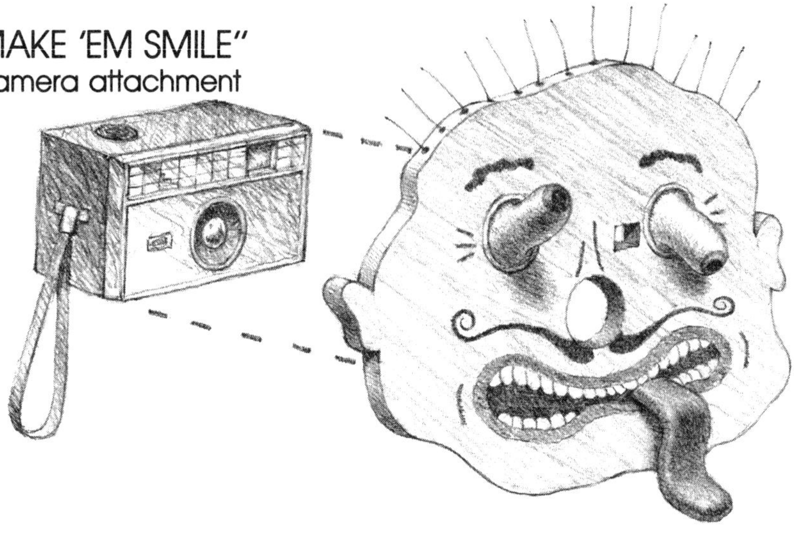

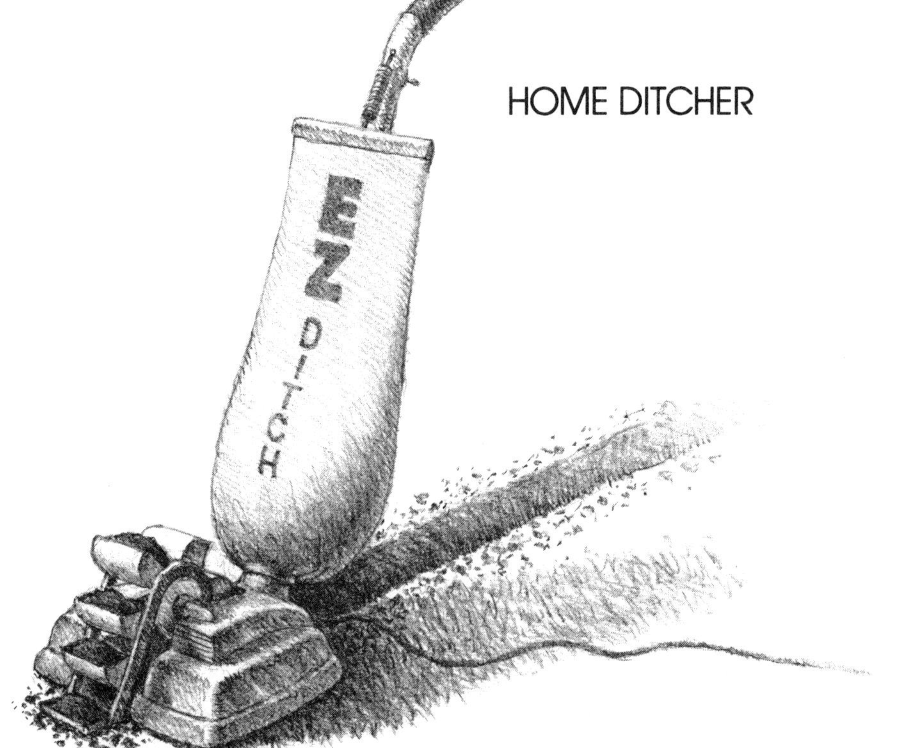

BODY SAILING

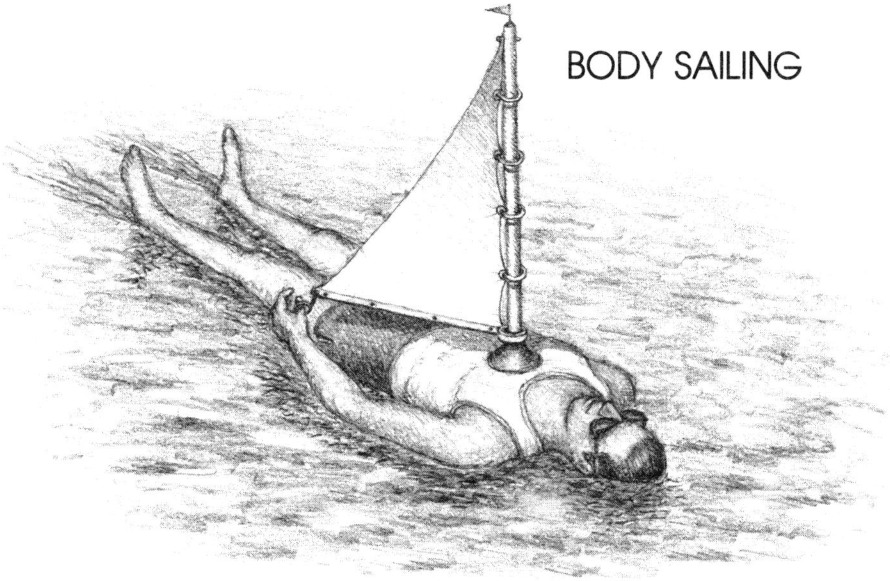

EARTHQUAKE SIMULATOR

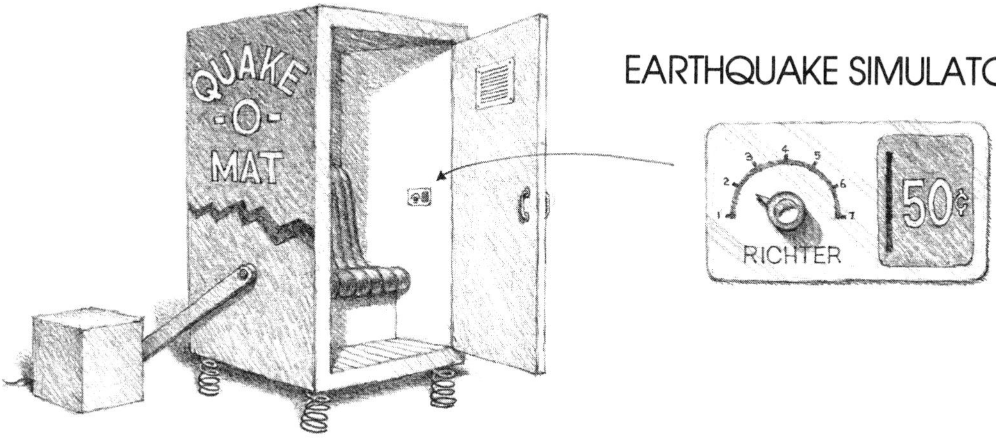

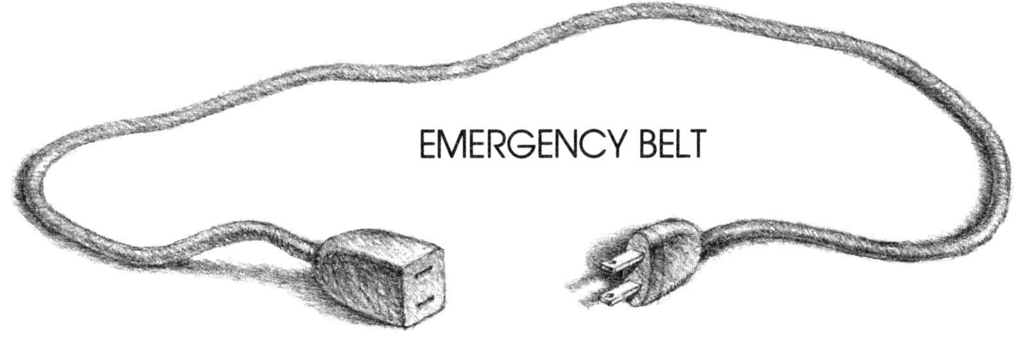

EMERGENCY BELT

"GOOD" MISSILES
cultural exchange, races, international practical jokes, etc.

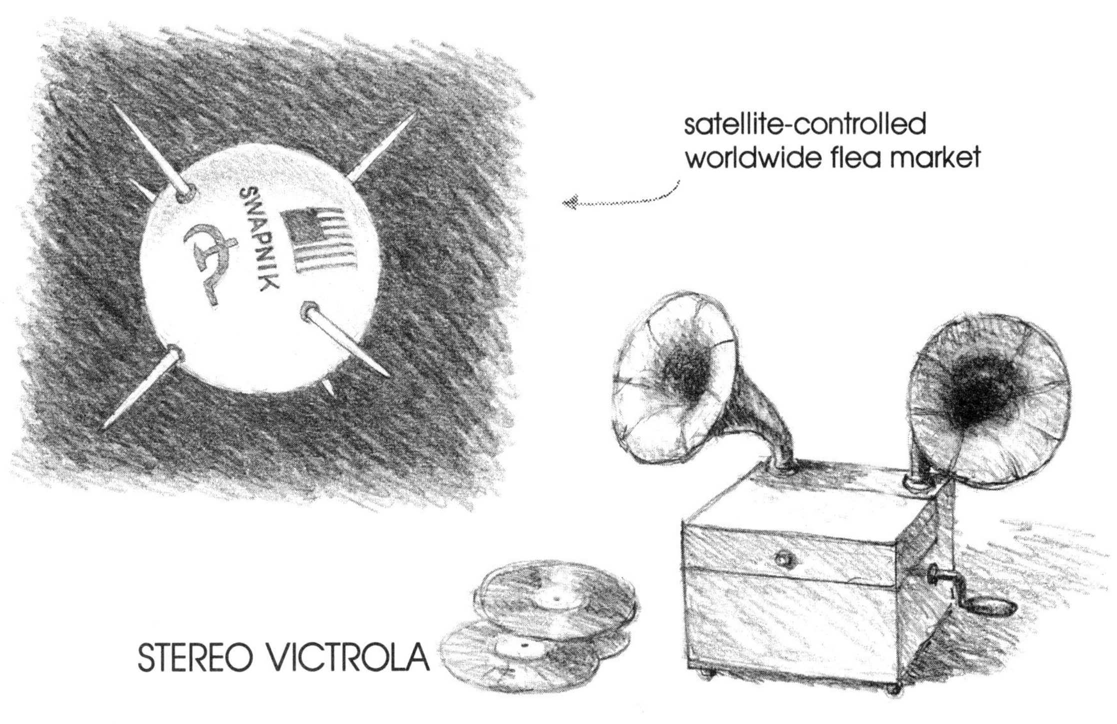

STEREO VICTROLA

AROMATIC

ROADSIDE SERVICE MODULE

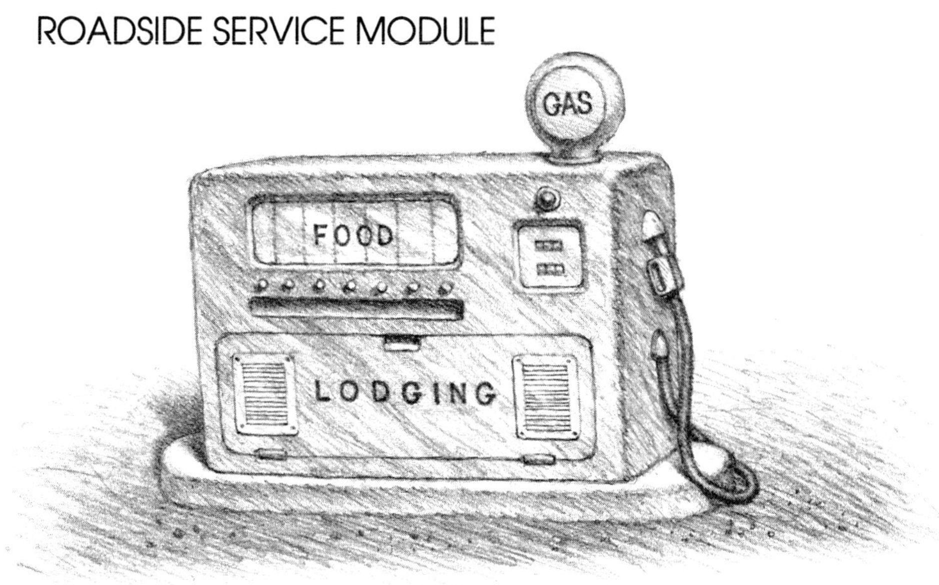

VIBRA-PEN

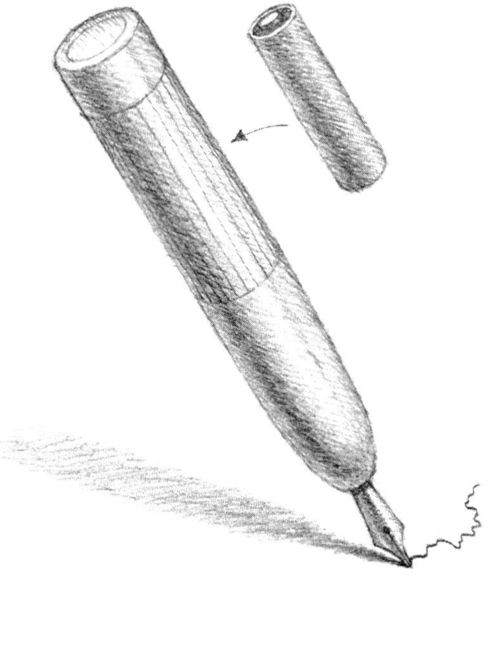

example

SELF-SHELF
manhole
NOVELTY "BIRDHOUSE" DOOR MOTIF
SEWER SUB

"FAT CHECK" DOORS

semi-sun

surf sound

"astro-sand"

SIMU-BEACH

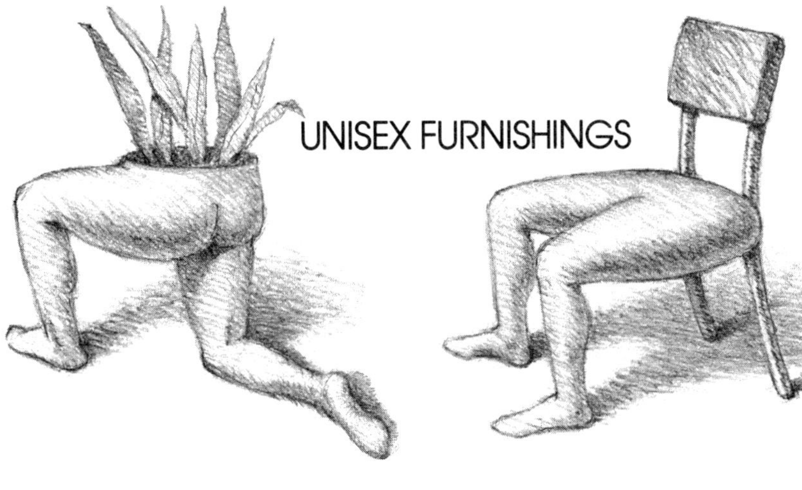

UNISEX FURNISHINGS

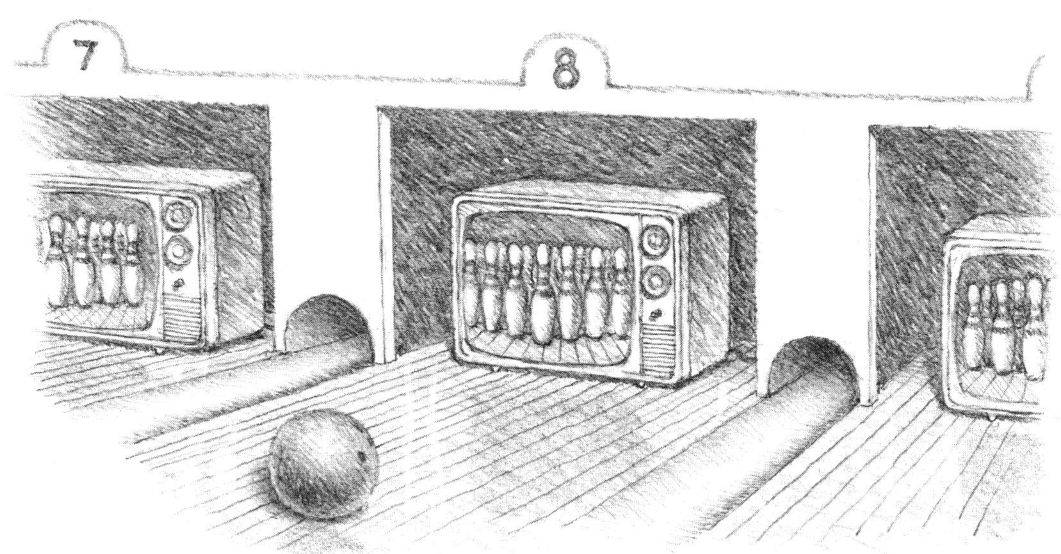

VIDEO BOWLING

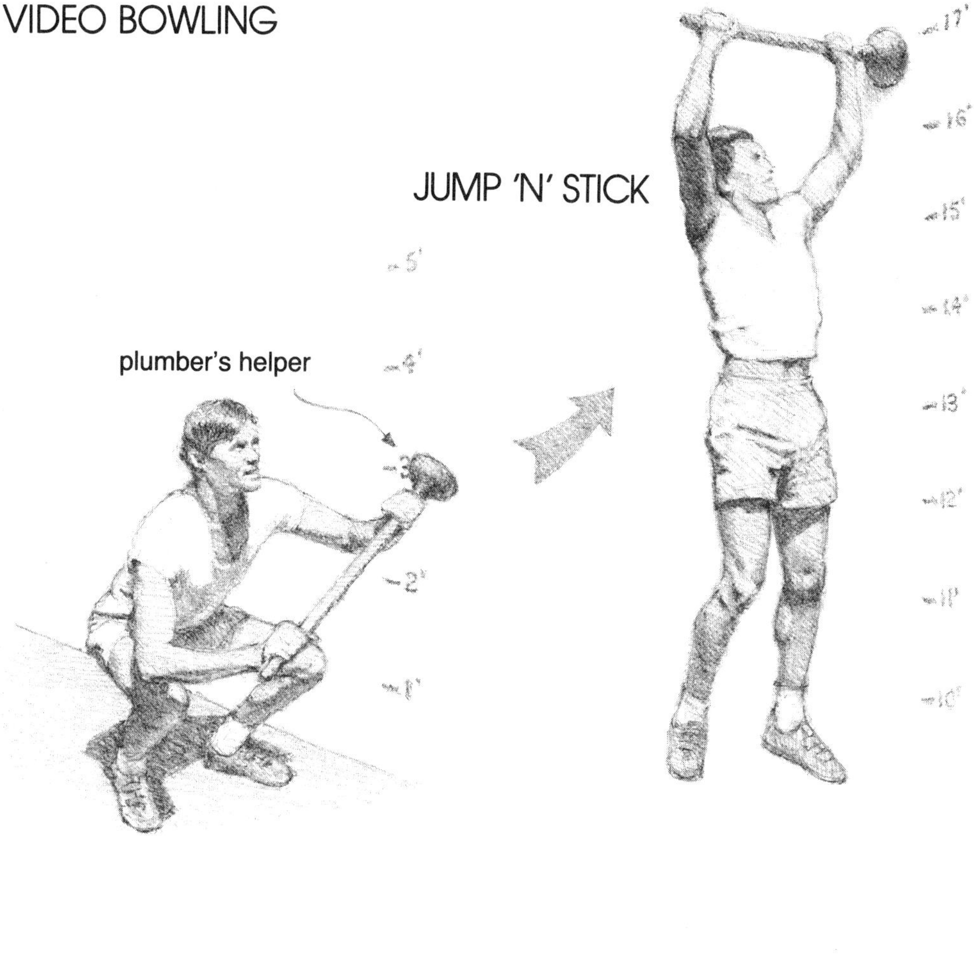

JUMP 'N' STICK

plumber's helper

POCKET-BELT (for nudists)

FLANGE SKIRT

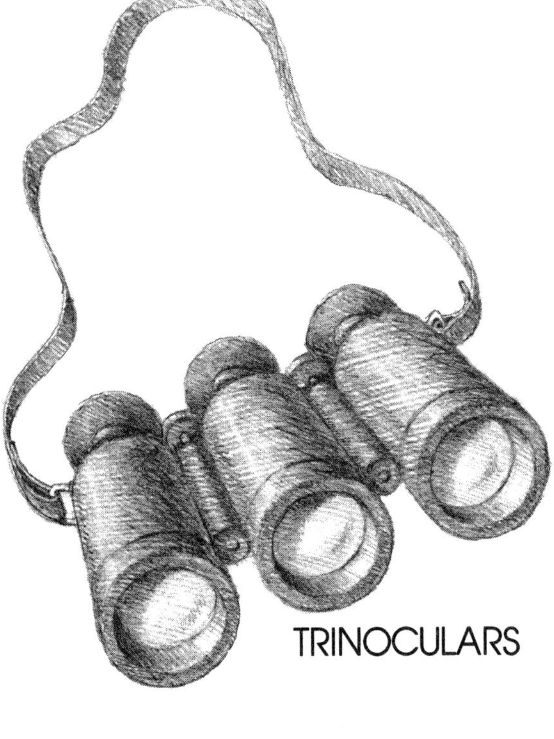

TRINOCULARS

"SUMPTY PUMPTY"

JUKE BOX FOOD SERVER

CEILING FIXTURE

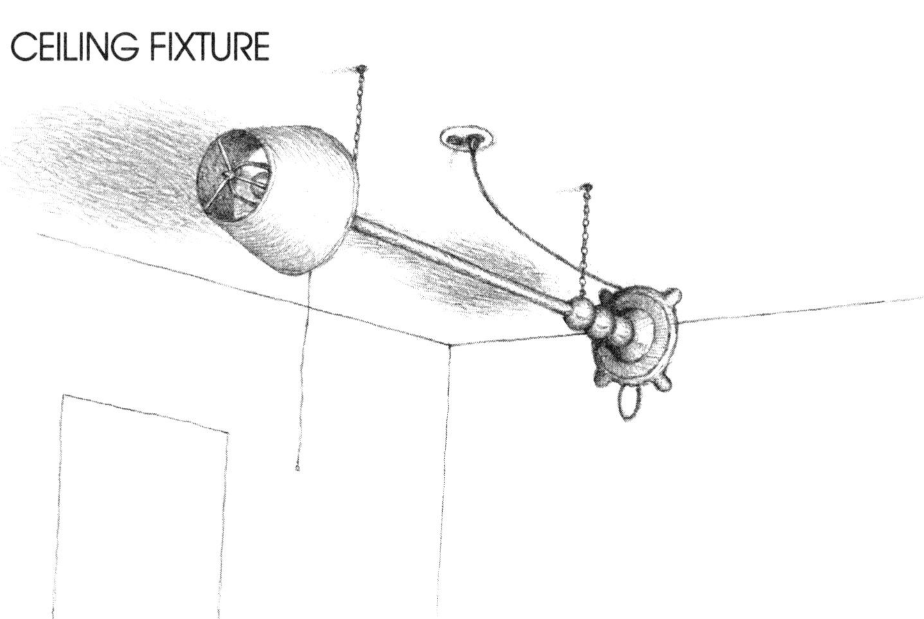

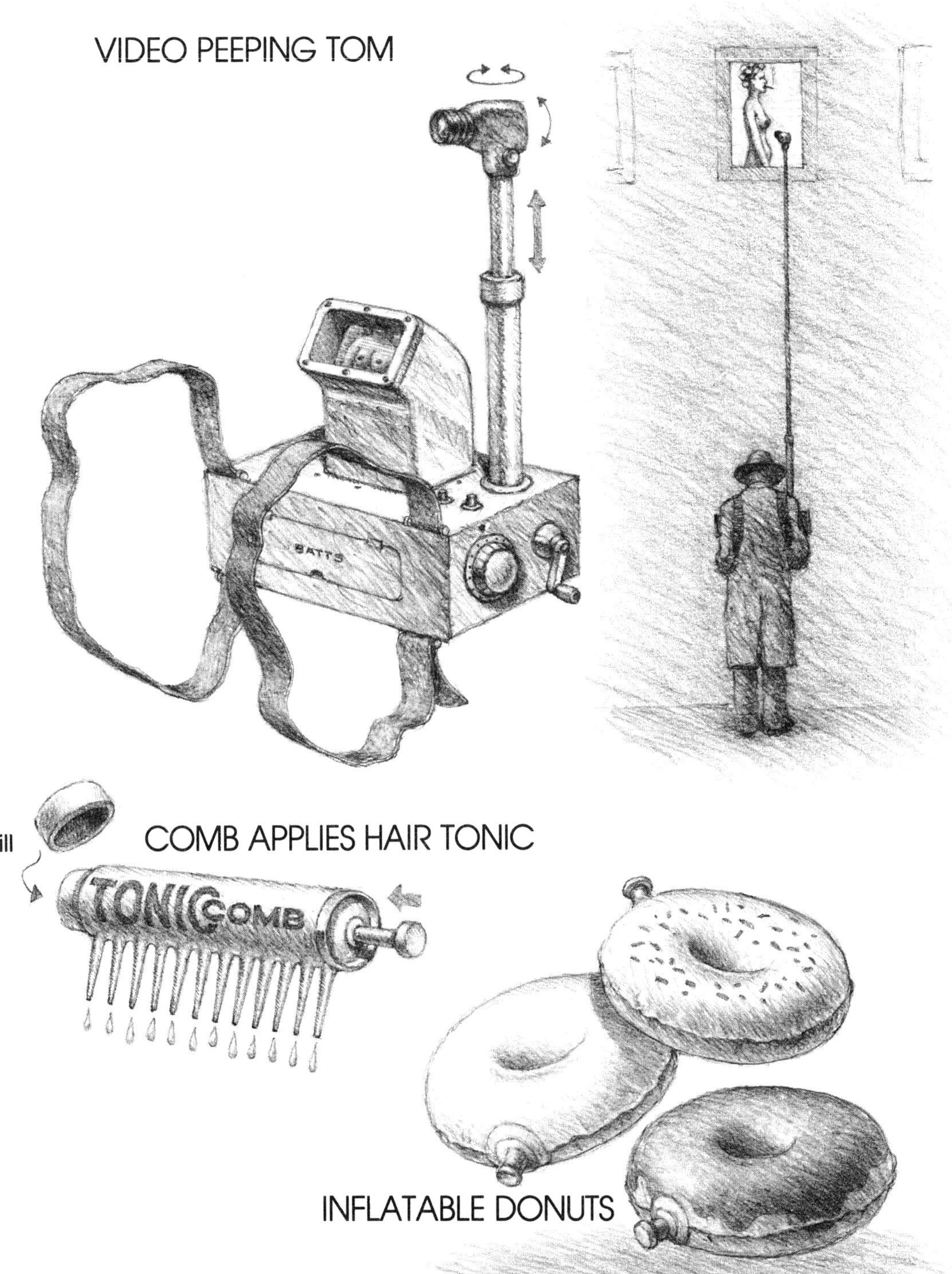
VIDEO PEEPING TOM
BATTS
fill
COMB APPLIES HAIR TONIC
TONIC COMB
INFLATABLE DONUTS

3-WHEEL MINI CAR made from rear of station wagon

COMMUNIST BLOCKS

PEEP-CYCLE

"DECENT" EXPOSURE TRUNKS

for legal exhibitionism

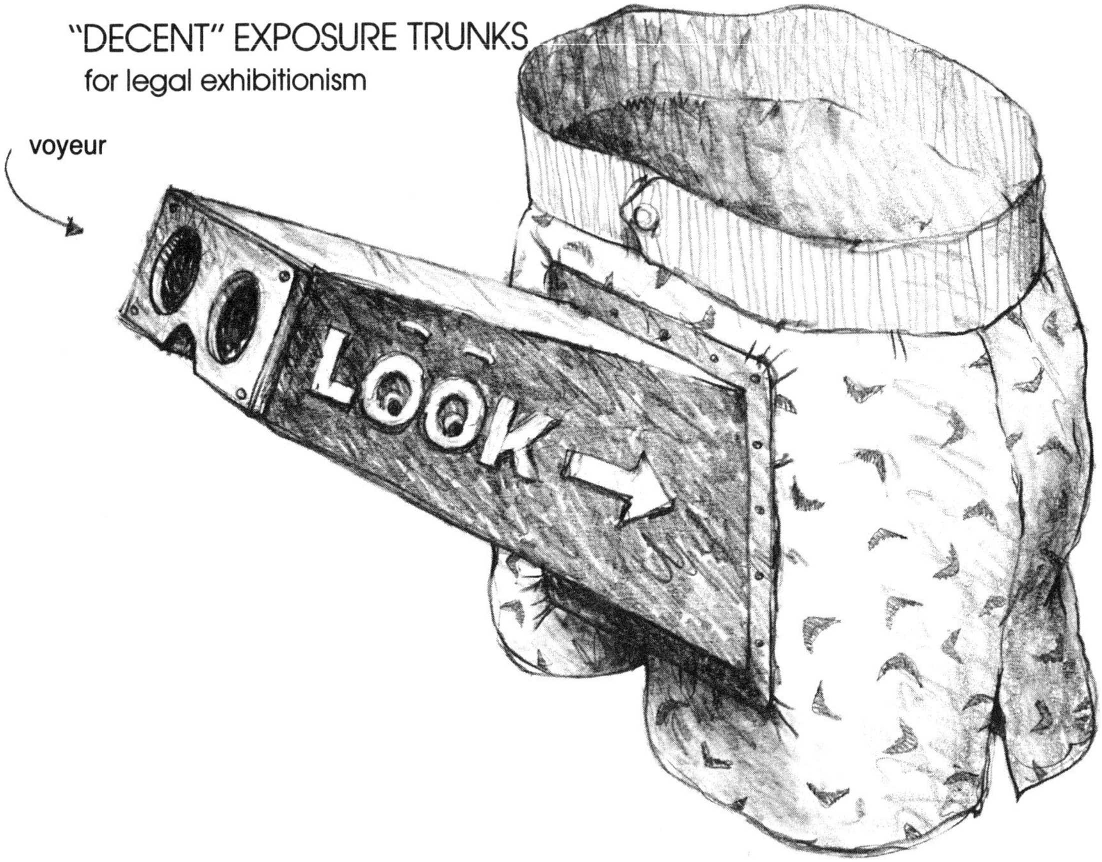

voyeur

LOOK

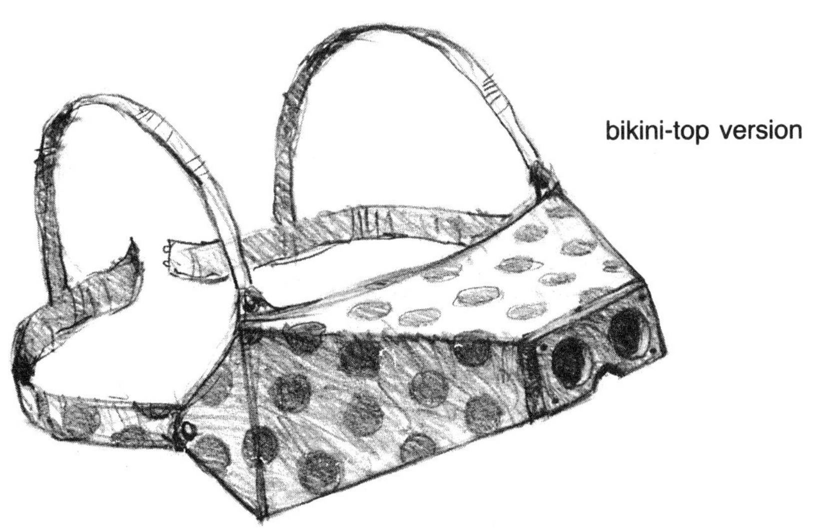

bikini-top version

"ONE-ON-ONE" INTRASENSORY COMMUNICATIONS SYSTEM

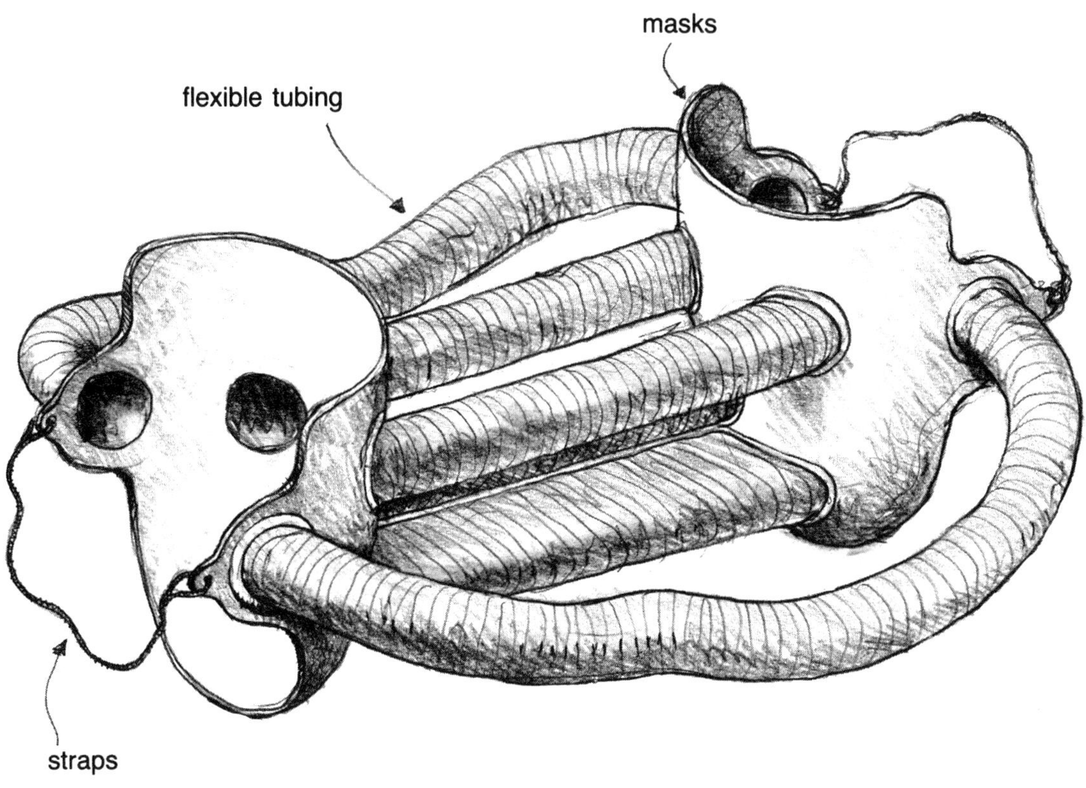

note: can be constructed
from halloween masks
and vacuum cleaner hose

GUITIRE

HUMAN DASHBOARD

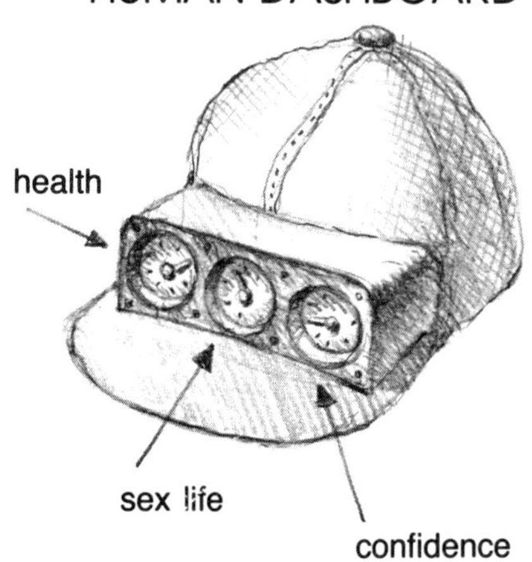

PHOTUM POLE

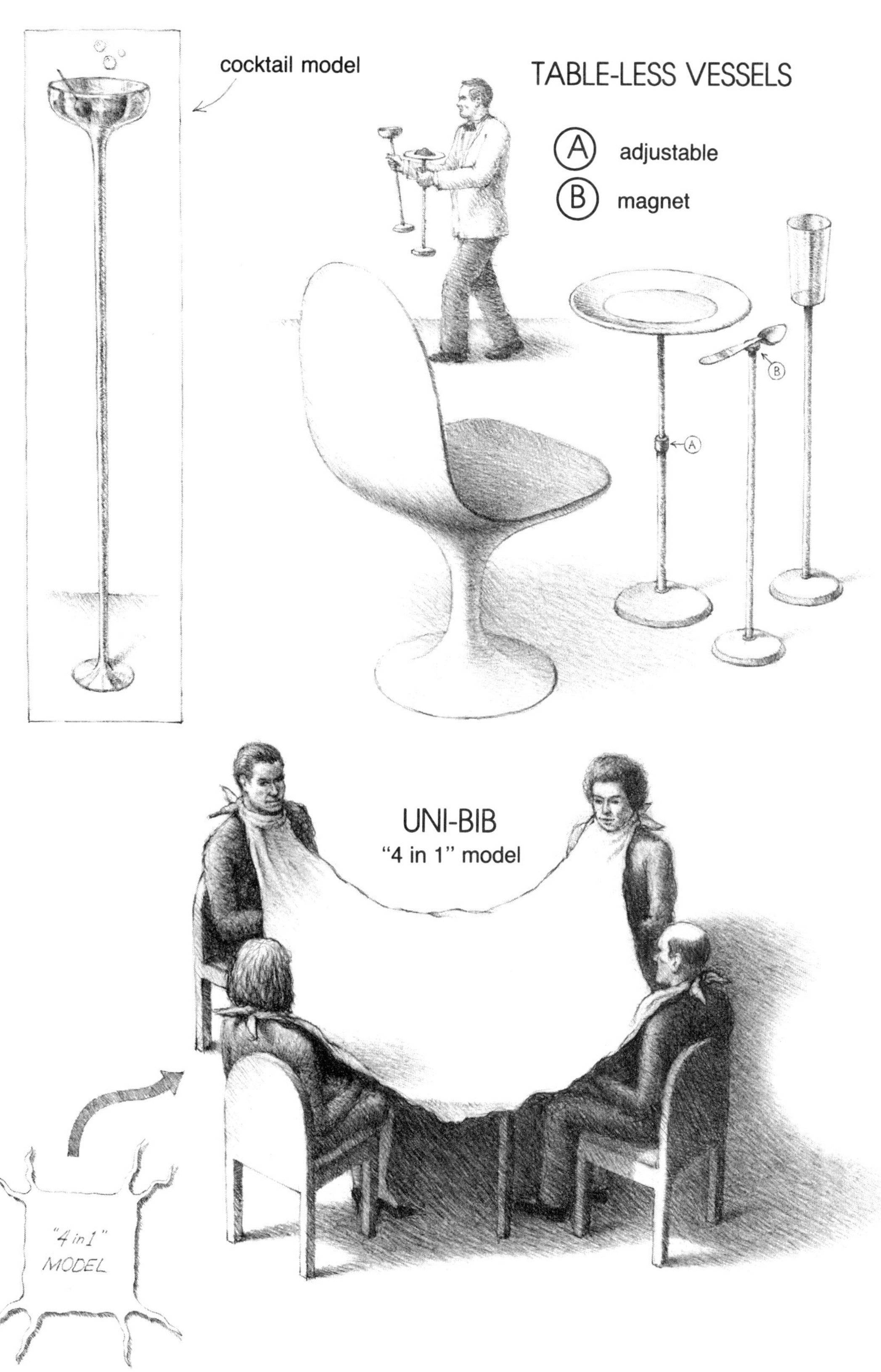
cocktail model
TABLE-LESS VESSELS
A adjustable
B magnet
B
A
UNI-BIB
"4 in 1" model
"4 in 1"
MODEL

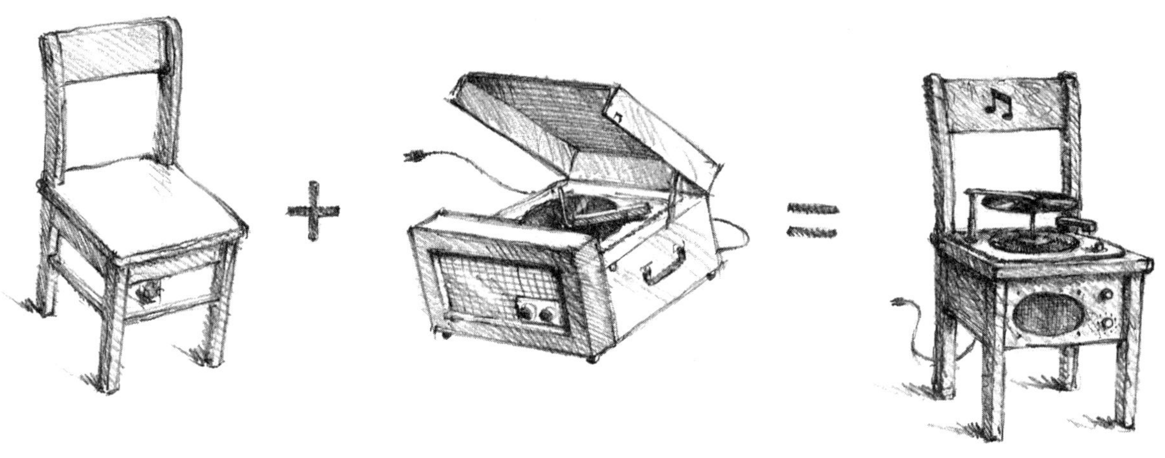

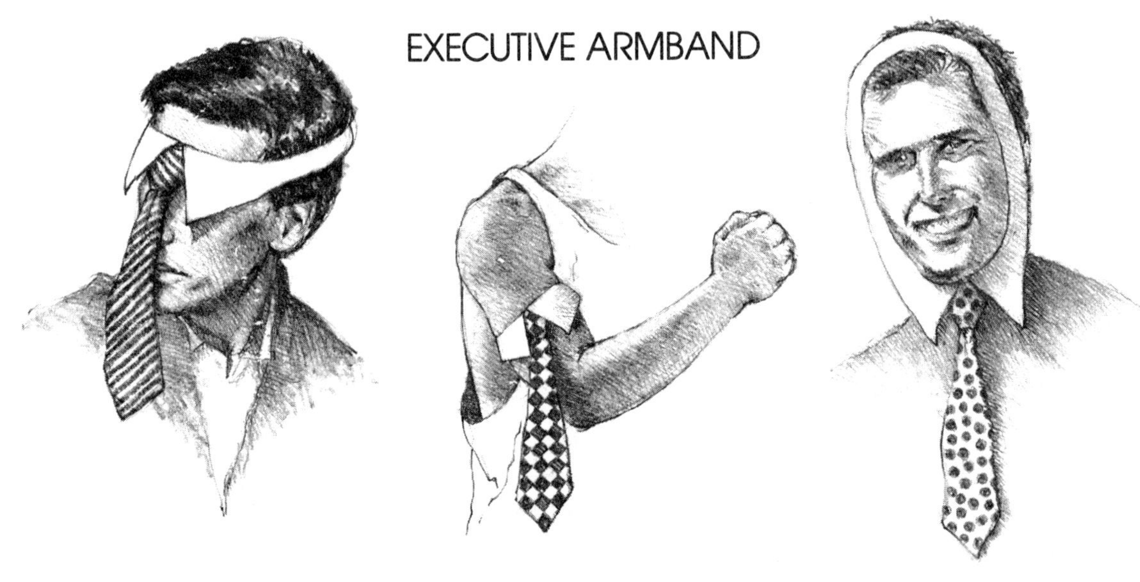
EXECUTIVE ARMBAND